Type your username and password or register by clicking on **Create a new account**.

Username

Password

In MyBook you can access the accompanying resources (both text and multimedia), the **BookRoom**, the **EasyBook** app and your purchased books.

CODE

> 1456Hhdo90

Type the code in the **Activation Code** field.

MY PURCHASED BOOKS

ENTER YOUR CODE — Activation code

The code must be typed only the first time you access **MyBook** and cannot be used thereafter.

CRITICAL THINKING

An Introduction

Edited by
Canale · Ciuni
Frigerio · Tuzet

Typesetting: Alberto Bellanti, Milano

Copyright © 2021 EGEA S.p.A.
Via Salasco, 5 - 20136 Milano
Tel. 02-58365751 - Fax 02-58365753
egea.edizioni@unibocconi.it - www.egeaeditore.it

All rights reserved, including but not limited to translation, total or partial adaptation, reproduction, and communication to the public by any means on any media (including microfilms, films, photocopies, electronic or digital media), as well as electronic information storage and retrieval systems. For more information or permission to use material from this text, see the website www.egeaeditore.it

Given the characteristics of Internet, the publisher is not responsible for any changes of address and contents of the websites mentioned.

First edition: August 2021

ISBN Domestic Edition 978-88-99902-85-8
ISBN International Edition 978-88-31322-49-2
ISBN Digital International Edition 978-88-31322-50-8

Print: Logo s.r.l., Borgoricco (PD)

Table of contents

Introduction — XI

Part I
Arguments and rationality

1 Rationality and cognitive biases — 3
- 1.1 Normative and descriptive approaches to reasoning — 3
  - 1.1.1 Critical Thinking, normative and descriptive — 3
- 1.2 Heuristics and biases — 7
  - 1.2.1 Framing effect — 9
  - 1.2.2 Hindsight bias — 14
  - 1.2.3 Anchoring effect — 18
- 1.3 The value of normative standards — 20
- 1.4 Conclusion — 22

2 What is an argument? — 23
- 2.1 Introduction — 23
- 2.2 Arguments and argumentative structure — 24
  - 2.2.1 Simple and complex arguments — 27
- 2.3 Good arguments vs. bad arguments — 30
  - 2.3.1 Deductive reasoning — 32
  - 2.3.2 Good arguments and deductively valid arguments — 35
  - 2.3.3 Kinds of non-deductive reasoning — 37
  - 2.3.4 Strength and weakness of non-deductive reasoning — 40
- 2.4 Conclusion — 41

3 Rational discussion and the pyramid of disagreement — 43
- 3.1 Introduction — 43
- 3.2 Rational discussion — 44
  - 3.2.1 The rules of rational discussion — 45
- 3.3 Ways of reacting to disagreement — 47
- 3.4 Conclusion — 55

4 How to reply rationally to an argument — 57
- 4.1 The two strategies for attacking an argument — 57
  - 4.1.1 Truth, justifiability and plausibility — 58
- 4.2 Attack on the argumentative structure — 60
  - 4.2.1 Non-sequitur — 61
  - 4.2.2 A fallacy of inductive reasoning: the unwarranted generalization — 62
  - 4.2.3 Circular reasoning — 63
- 4.3 Attack on the premises — 66
  - 4.3.1 Counterexamples — 67
  - 4.3.2 Request for justification — 67
  - 4.3.3 Reductio ad absurdum — 68
- 4.4 Conclusion — 69

Part II
Deductive arguments

5 Deductive arguments — 73
- 5.1 Introduction — 73
  - 5.1.1 Semantics and syntax. Classical logic — 77
- 5.2 Propositional reasoning — 79
  - 5.2.1 The truth conditions of complex sentences — 82
  - 5.2.2 Definability — 86
  - 5.2.3 Proving by rules. Natural deduction — 92
  - 5.2.4 Assertion and rules of inference — 97
- 5.3 Reasoning with predicates and quantifiers — 99
  - 5.3.1 Sentences containing predicates and quantifiers, and their relationships — 101
  - 5.3.2 A formal language for reasoning with predicates and quantifiers — 105
  - 5.3.3 Some remarks on the logical regimentation of predicates and quantifiers — 113
  - 5.3.4 Categorical syllogisms — 114
- 5.4 Conclusion — 117

6 Conditional reasoning, I: The material conditional — 119
- 6.1 Introduction — 119
- 6.2 A reasoning problem involving conditionals — 121
- 6.3 The conditional — 123

		6.3.1 The many kinds of indicative conditional	124
		6.3.2 Truth and falsity of indicative conditionals	125
	6.4	The material conditional	127
		6.4.1 The paradoxes of material implication	129
		6.4.2 Inference rules with material conditionals	133
		6.4.3 Two fallacies and other rules of inference	134
	6.5	Conditionals and necessary and sufficient conditions	137
	6.6	Other views of indicative conditionals	140
	6.7	An experiment involving modus tollens	142
	6.8	Conclusion	145
7	**Conditional reasoning, II: The counterfactual conditional**		**147**
	7.1	Introduction	147
	7.2	The psychology of counterfactual reasoning	150
		7.2.1 The importance of counterfactual reasoning	150
		7.2.2 The fault lines: which aspects of reality do we modify?	151
	7.3	The logic of counterfactuals	154
		7.3.1 Indicative and counterfactual conditionals	156
		7.3.2 The truth conditions of counterfactuals	157
		7.3.3 The Ramsey Test for counterfactuals	158
		7.3.4 The semantics of counterfactuals	160
		7.3.5 Fallacies of counterfactual reasoning	166
	7.4	Counterfactuals and causality	172
		7.4.1 The counterfactual test	173
		7.4.2 Inadequacy of the counterfactual test	174
	7.5	Conclusion	175

Part III
Non-deductive arguments

8	**Reasoning with explanatory hypotheses**		**179**
	8.1	Introduction	179
	8.2	A medical case	180
	8.3	Abductive procedures	183
		8.3.1 Abduction	183
		8.3.2 Inference to the best explanation	185
	8.4	An example of reasoning with explanatory hypotheses: reasoning from effects to causes	188
		8.4.1 The method of agreement	191
		8.4.2 The method of difference	192

		8.4.3 The joint method of agreement and difference	193
		8.4.4 A more complex case	194
		8.4.5 An assessment of Mill's methods	195
	8.5	Conclusion	196

9 Statistical reasoning — 197

	9.1	Introduction	197
	9.2	Reasoning with purely statistical generalizations	198
		9.2.1 Lack of relationship between the target and measured property	199
		9.2.2 Cardinality of the sample	200
		9.2.3 Biased samples	201
		9.2.4 Question biases	203
		9.2.5 Hidden variables	204
		9.2.6 Some considerations on the margin of error	204
		9.2.7 Statistical generalizations and daily life	205
	9.3	Generalizations concerning the future	205
	9.4	Statistical syllogism	208
		9.4.1 Statistical syllogisms and reference classes	210
	9.5	Conclusion	212

10 Probability and probability biases — 213

	10.1	Introduction	213
	10.2	Three interpretations of probability	215
		10.2.1 The classical interpretation	216
		10.2.2 The frequentist interpretation	217
		10.2.3 The subjective interpretation	218
	10.3	Basic aspects of probability theory	220
		10.3.1 Rules for intersection	223
		10.3.2 Rules for union	224
		10.3.3 Rule for complement event	225
		10.3.4 Bayes' theorem	226
	10.4	Two probabilistic reasoning problems, and some fallacies	230
		10.4.1 The cab problem	232
		10.4.2 The prosecutor's fallacy	233
	10.5	Risk and fundamental uncertainty	234
		10.5.1 Risk	234
		10.5.2 Fundamental uncertainty	235
	10.6	Conclusion	237

11 Reasoning by analogy — 239
- 11.1 Introduction — 239
- 11.2 Structure of arguments by analogy and evaluation criteria — 240
- 11.3 Reasoning by analogy with cognitive purposes — 241
- 11.4 Analogy in legal reasoning — 242
- 11.5 Analogy in moral reasoning — 245
- 11.6 Conclusion — 246

Further reading — 247

Introduction

What you are holding in your hands is an introduction to Critical Thinking. This discipline regards two related areas: first, the set of skills and competences that help you recognize, evaluate and produce *good* arguments; second, the set of activities and tools that we can call on and use to train these skills, respectively. This being so, why is the subject called "Critical Thinking"?

In ancient Greece, the term "kriticós" meant the ability to judge and discern things, in order to make the best choices for ourselves and others. These days, this ability appears to be struggling to survive. In the world of hyper-communication, monopolized by social media, it happens that our ability to judge the opinions of others, to tell truth and falsity apart, to distinguish good reasons from bad reasons in support of a choice, has been weakened almost to the point of disappearing. We all know why. The inflated amount of information showering us every day makes it very hard to tell what is reliable and relevant from what is not. This phenomenon is just made worse by the proliferation of *fake news*, that is to say by the deliberate dissemination of false information that leverages widespread prejudices plus our fears, thus managing to attract the attention of the public. In addition, today any exchange of views, both on the social media and in public discussion, tends to turn into a relentless struggle, in which the insult, the denigration of the opponent, or the ridiculing of the opinion of others, often takes the place of critical discussion of the reasons supporting a given claim. All this drastically reduces not only the quality of public debate and collective choices, but also our ability to express a well-considered judgment on issues that are often very relevant to our lives.

It is therefore no coincidence that today the graduation programs of many of the most important universities in the world include a Critical Thinking course: it aims at sharpening the students' ability to adequately justify a claim, to refute the claims of others, to identify mistakes in reasoning, and to evaluate the reasons in support of a certain assertion. This applies both to everyday discourse and scientific inquiry.

The goal in question can only be achieved, however, by adopting an interdisciplinary approach. There is a vast spectrum of disciplines from which one can draw in this respect: logic, probability theory, statistics, decision theory, and again the theory of argumentation and the theory of rational discussion, as well as disciplines which study language, first of all pragmatics. An important contribution to a full awareness of the problems and skills that come into play in Critical Thinking is also provided by cognitive psychology, and in particular the psychology of reasoning and the psychology of decision. The ability to produce and evaluate arguments, in fact, depends not just on the rules that govern rational discussion, but also on our ability to guard against cognitive biases that systematically lead us into error in making a choice or in providing a solution to a problem. More in general, the ability in question depends on our ability to avoid fallacies, that is, patterns of reasoning that do not work, and yet we tend to follow them for some reason. These mistakes are due to how *real cognitive agents* are made – to our difficulties in using some reasoning patterns; or to the fact that the cognitive fatigue that we would experience in following correct reasoning procedures leads us to adopt "shortcuts" that can lead us to wrong conclusions.

One important point of Critical Thinking is that, in teaching us to evaluate an argument, it does not take *the way in which we actually tend to think* as a criterion. Instead, it assumes a set of *normative standards* by which *we must* reason, *if* we want to be *rational*, as is desirable. Indeed, the Critical Thinking approach considers reasoning from a normative point of view, not from a *descriptive* one. It is worth noticing that we ourselves recognize the normative prescriptions on reasoning as rational once we are told where our mistakes are, as we will see in Chapter 1. We will see, in any case, that a normative approach to reasoning can only really be illuminating if we are aware of how we *actually* reason. Thanks to cognitive psychology, today we can achieve this (at least, to a much greater degree than in the past).

Being aware of the various aspects that condition the reliability of the judgments in each field of knowledge will help the student to acquire the critical and anti-dogmatic spirit that distinguishes a good university education. Moreover, learning to think critically is fundamental today for anyone wishing to make thoughtful choices in the field of public life and to avoid the temptation to believe that the opinion of those who shout the loudest, or receive the most likes, is necessarily the best option. In this sense, the path proposed in these pages is aimed not only at university students, but at anyone who has the desire to learn to think better and discover the weaknesses of other people's reasoning.

Contents and structure of the volume

This book is an introduction to Critical Thinking, which first focuses on its logical aspects, showing at the same time the intersection of logic with the study of cognitive biases and also with the characteristics of the argumentative processes that have a place in everyday life, without neglecting an introduction to the basic tools for probabilistic and statistical reasoning. The path proposed in the book is divided into three parts.

In the first part (Chapters 1-4) we will focus on what a *good* argument is, and we will explore the function that the evaluation and exchange of good arguments play in our rational activities. We will do this after having given a context in which we can frame the activity of evaluating and producing good arguments. More specifically, Chapter 1 will deal in detail with the distinction between *normative approaches* and *descriptive approaches* to reasoning, allowing us to understand that there are standards of *rationality* that, for a variety of reasons, we do not always follow. Chapter 2 will focus more specifically on what *arguments* and *good arguments* are, thereby elucidating two concepts that we will use throughout the rest of the volume. The second – that of a good argument – is the key concept of this work. We will see, in the same chapter, that there are different forms of reasoning. Chapter 3 will set arguments and their features in a precise context, that of discussion, and in particular rational exchanges of views, and will address a phenomenon that always occurs in the most interesting discussions, namely disagreement. Indeed, it is the presence of disagreement which solicits the elaboration of arguments. We will give an overview of the different reactions to disagreement, concluding with the most rational one. Chapter 4 will analyze the latter in detail, illustrating the different strategies we can use to concretely realize that.

This first part offers a fundamental set of notions that constitute the common thread of the volume, and provides a general framework that helps us understand why we should take interest in reasoning, and that reasoning is fundamental in many applicative disciplines and in many concrete areas of our lives.

In the second part (Chapters 5-7) we will discuss deductive arguments, which are one of the most solid and best regimented tools of our reasoning activities. Chapter 5 presents an overview of the reasoning tools provided by propositional (or "sentential") logic and quantified (or "predicate") logic. It also shows concrete strategies for checking whether an argument is deductively valid or not (these notions will be introduced in Chapter 2). Chapter 6 and Chapter 7 delve into two specific cases of deductive reasoning: reasoning with indicative conditionals, and reasoning with counterfactuals. The reason for dealing with them is that they are both central to our reasoning activities. Chapter 6 focuses mainly on a particular type of indicative conditional that logicians refer to as a "material conditional", but it also discusses the possibility of other indicative conditionals.

That is the most technical part of the volume, even if the material has been conceived and written to be accessible to an audience with no background in logic or mathematics. The technical aspects are explained step by step, keeping notation and formal considerations to a minimum.

In the third part (Chapters 8-11) we will discuss non-deductive arguments. However solid deductive reasoning is, it does not, in fact, prove particularly useful or illuminating in some contexts, especially those in which we try to explain facts by formulating *hypotheses*, or where we have to reason about *statistical projections* or *probabilities*, or in contexts in which our reasoning must be conducted on the basis of *analogies*. Chapter 8 deals with reasoning using explanatory hypotheses, and discusses a specific case: reasoning with causal hypotheses. Chapter 9 deals with statistical reasoning – both with *statistical generalization* and with the *application of statistics* to specific cases. Chapter 10 deals with probabilistic reasoning. It is the most technical chapter of the third part, but the formal and mathematical considerations are still kept to a minimum. Finally, Chapter 11 deals with reasoning by analogy, which plays a considerable role in legal and moral reasoning.

This last part is more discursive than the second, as we have favored conceptual understanding over technical discussion (keep in mind, however, what we have just said about Chapter 10). This is because there are a great many formal approaches to the various types of non-deductive reasoning, while there is much more uniformity as regards deductive reasoning. Since this is an introduction to Critical Thinking, we have favored the presentation and understanding of the basic aspects over a technical study that could branch out in many directions, with the risk of going far beyond the didactic purposes that the book sets out to achieve.

Although the discussion follows a unitary line of development, the various chapters lend themselves to being read and used separately, according to the needs of the readers and their desire for further study.

It is not entirely possible to fully understand, evaluate and produce good arguments if we do not familiarize ourselves with the "other side of the coin", that is, with those reasoning patterns that seem to work but in reality systematically lead us to erroneous conclusions. This is one of the reasons why we discuss *cognitive biases* in Chapter 1. Furthermore, we discuss some fallacies where these seemed most relevant. Chapter 4, Chapters 6-7 and Chapter 10 discuss specific fallacies and errors of reasoning in the corresponding areas of reasoning covered by the chapter.

The volume tries to follow a precise methodology of presentation: we start out from examples as far as possible and use these to illustrate the general and conceptual points we face. We believe that a bottom-up treatment is more suitable for addressing specific problems than a top-down one, which favors the systematic presentation of theory and general considerations over case studies. However, we have tried to ensure that the exemplifications do not diminish the attention paid to those general and abstract aspects that are necessary for the conceptual understanding of reasoning and of the problems connected to it.

A terminological note

Some of the terminological choices we made in writing this volume might sound misleading to the reader who is already familiar with philosophy or the theory of argumentation, or with the theory of reasoning in general. We therefore believe that it is appropriate to discuss them briefly.

Throughout the volume we will talk about *sentences* rather than *propositions*. We will say that sentences appear in reasoning and that sentences have truth-values. An important philosophical tradition, however, has it that propositions do this. What is the difference? From a technical point of view, a sentence is a syntactic entity that (i) satisfies some rules of grammar (or is "well formed", as logicians say), and that (ii) if uttered in the indicative mood asserts something. "La neve è bianca" is a sentence in the Italian language. "Snow is white" is a sentence in the English language. "Umberto Eco" is *not* a sentence, but a name – if I utter this string of sounds I am not asserting something, just pronouncing a couple of words. Since the identity criterion for sentences is their syntactic make-up, "La neve è bianca" and "Snow is white" are different sentences. Instead, a *proposition* is the *content* of a sentence. But there is some disagreement as to what a proposition is, in turn. For some philosophers, a proposition is the thought expressed by a sentence; for others, it is (ideally) the set of scenarios that make a given sentence true. Be that as it may, both views agree in saying one crucial thing about propositions, as the example shows: "La neve è bianca" and "Snow is white" are different sentences that express the *same* proposition. More generally, we could say that propositions are "interpreted sentences", that is, the result of the association of a content (be it a thought or a set of scenarios) to a string of signs having certain characteristics. From this point of view, it is perfectly justified to say that it is the propositions that are true or false. In fact, it is by virtue of "interpretation", in the sense just illustrated, that they are true or false. And it is perfectly justified to say that reasoning is made up of propositions. Reasonings are mental acts (acts that involve certain entities: the arguments), and therefore involve sentences that we always *think of* or, in a broader sense, that we "interpret", in the sense illustrated.

However, we believe that relaxing the terminology and saying that a sentence has a truth-value will do no harm. In fact, given *one* particular interpretation of it, a sentence can be associated with truth-values *by transfer*, on the basis of the truth or falsity of the proposition which that given interpretation associates with it. If so, we can also, for the sake of convenience, take the sentences to be constituent parts of arguments and reasonings, when, as in this volume, the main interest is in the truth of the premises and the conclusion, and of their logical relationships. Indeed, these relationships can be faithfully transferred from propositions to sentences.

Let us therefore take this terminological liberty, in the awareness there are certainly grounds for suggesting that the other choice would be more rigorous. One reason we were prompted to proceed in this way is that, given the disagreement on what a proposition is, it is easier and more intuitive to explain and understand what a sentence is. In any case, our choice is applied consistently throughout the book.

Another clarification: in this volume we talk about *arguments*, *reasoning*, and *inferences*. We will see in Chapter 2 what an argument is. Technically, reasoning is the act – carried out by any cognitive agent – of *thinking about* a topic, that is, of carrying it out in thought; in this sense, reasoning is a *mental act*. However, in history and even in current usage, the term "reasoning" is almost always used in a looser sense, and in such a way that it is in fact interchangeable with the term "argument". We will do the same throughout this volume. This choice will do no harm: as you will notice when reading the text, nothing we say when talking about "reasoning" will depend on its specific nature as a mental act – everything will instead depend on the "structural" characteristics that reasoning "inherits" from the relative argument. We will therefore speak, for example, of the validity or invalidity of an instance of reasoning (or an argument), and of the structure of an instance of reasoning (or an argument). Finally, inference is the act of concluding a sentence (or proposition) from a series of other sentences (or propositions) in a reasoning. From a technical point of view, therefore, it is another mental act. However, in this case too, the use of the term is usually looser. In this volume, we will talk about inference whenever we are in the presence of the particular application of an argumentation scheme that leads us from some premises to a conclusion, regardless of whether the premises and conclusion are thought of or not, and whether the act of passing from the premises to the conclusion is actually accomplished. Consistently with this choice, we will use "inference scheme", "reasoning scheme", and "argumentation scheme" as synonyms.

Origin and attribution of the chapters

It is important to point out that this book is the result of a collective work. We started thinking about it in 2018, when the first Critical Thinking course was given at Bocconi, at the initiative of Rector Gianmario Verona. The course is still taught to a considerable number of classes, which required the collaboration of several teachers and authors. In the first phase, each chapter was assigned to a specific author; subsequently, the chapters were substantially reworked by Roberto Ciuni and Aldo Frigerio, and then revised by Damiano Canale and Giovanni Tuzet.

The initial versions of the chapters are attributable as follows: Chap. 1, Giovanni Tuzet; Chap. 2, Damiano Canale; Chaps. 3-4, Ciro De Florio and Aldo Frigerio; Chap. 5, Roberto Ciuni and Aldo Frigerio; Chap. 6, Massimiliano Carrara; Chap. 7, Vittorio Morato; Chap. 8, Roberto Ciuni and Giovanni Tuzet; Chap. 9, Aldo Frigerio; Chap. 10, Daniele Chiffi; Chap. 11, Aldo Frigerio and Giovanni Tuzet.

The final versions published in this book can be attributed as follows:

- Chapter 1, Roberto Ciuni, Aldo Frigerio, Giovanni Tuzet;
- Chapter 2, Damiano Canale, Roberto Ciuni, Aldo Frigerio;
- Chapters 3-4, Roberto Ciuni, Ciro De Florio, Aldo Frigerio;
- Chapter 5, Roberto Ciuni, Aldo Frigerio;
- Chapter 6, Massimiliano Carrara, Roberto Ciuni, Aldo Frigerio;
- Chapter 7, Roberto Ciuni, Aldo Frigerio, Vittorio Morato;
- Chapter 8, Roberto Ciuni, Aldo Frigerio, Giovanni Tuzet;
- Chapter 9, Roberto Ciuni, Aldo Frigerio;
- Chapter 10, Daniele Chiffi, Roberto Ciuni, Aldo Frigerio;
- Chapter 11, Roberto Ciuni, Aldo Frigerio, Giovanni Tuzet.

Finally, we wish to thank Elisabetta Lalumera and Alessio Sardo for their contribution to the first edition of the Bocconi course. But above all we should thank the students who, by practicing their Critical Thinking, helped us to improve the materials from which this book originated.

THE AUTHORS

Damiano Canale, Bocconi University, Milan
Massimiliano Carrara, University of Padua
Daniele Chiffi, Politecnico di Milano
Roberto Ciuni, Roma Tre University
Ciro De Florio, Catholic University of the Sacred Heart, Milan
Aldo Frigerio, Catholic University of the Sacred Heart, Milan
Vittorio Morato, University of Padua
Giovanni Tuzet, Bocconi University, Milan

Part I
Arguments and rationality

1 | Rationality and cognitive biases

In this chapter we distinguish between normative approaches and descriptive approaches to reasoning (§ 1.1), and discuss concerns regarding the method and empirical adequacy of normative approaches. We will explore three cognitive biases (§ 1.2) to see how we often fail to comply with the normative standards of reasoning. We will see that these examples help us understand why, despite the difficulties of the normative approaches, the standards of rationality they set are indispensable (§ 1.3).

1.1 Normative and descriptive approaches to reasoning

A fundamental component of Critical Thinking involves the analysis of our reasoning patterns – those we perform, those we hear in a public debate of any kind, those that are proposed to us by someone else, etc. When it comes to analyzing our reasoning, there are two major dimensions we may be interested in. One is the *normative* dimension, which allows us to identify some *standards* of *correctness* and *rationality* for the reasoning we carry out (for example) in proposing an argument to someone, in solving a mathematical problem, or in making a decision. The other is the *descriptive* dimension, which concerns how actual cognitive agents propose arguments, tackle a mathematical problem and make decisions. There are disciplines which mainly address our rational cognitive activities with normative purposes, and disciplines which study reasoning with descriptive purposes. We will give some examples of them shortly. Meanwhile, we can say, in a few words, that the *normative* disciplines tell us how we should reason in order to reason *correctly*, while the *descriptive* ones tell us how we *actually* reason.

The central part of this chapter will show how the two do not necessarily go hand in hand: we do not always think correctly. In particular, we will see that we tend to fall victim to some patterns of reasoning that we ourselves come to consider incorrect, once they are pointed out to us. The examples we will consider in this chapter involve so-called *cognitive biases*. We will see shortly what these are.

1.1.1 Critical Thinking, normative and descriptive

From what we said in the Introduction, it is clear that Critical Thinking deals with the normative correctness of our reasonings. In a nutshell, we can say that Critical Thinking is about *how to think*, not how we think. In fact, how we think is a matter of psychological facts, while how to think is a matter of normative correctness. As such, Critical Thinking intertwines to some extent with other normative disciplines involving our cognitive and reasoning activities. The most important of them are indicated in this list:

- logic;
- theory of probability;
- neoclassical economics;
- theory of rational choice;
- theory of decision;
- game theory.

Logic concerns the relation of logical consequence between (sequences of) sentences (in the case of deductive logic) and of the evidential support that some sentences provide to others (inductive and abductive logic).[1] More generally, we can say that logic deals with the rational connections between sentences. *Probability theory* deals with how to strictly define and mathematically measure the probability of an event. *Neoclassical economics* (and in particular the microeconomic theories that refer to it) defines economic agents as maximizers of their own utility, where the latter is constrained by factors and assumptions that in fact idealize the behavior of economic agents. The classical *theory of rational choice* sets standards so that the choices of agents are *rational* – we will see what this means, exactly. *Classical decision theory* provides standards of rationality for the decisions of agents, and *classical game theory* offers a package of analytical tools with which to investigate our strategic interactions, that is, those in which we have to consider not only our goals, but also the fact that others pursue their goals, and that what we get depends on the choices all agents involved make.

None of these disciplines deals directly with reasoning, except logic (but we will soon see that the relationships between logic and reasoning are more subtle than we usually think), although each of them has something to do with it indirectly. It is easy to understand why: to solve a problem of logic, I have to think; to solve a problem related to the probability of an event, I have to think based on the quantitative data I have (assuming I have the necessary data); to make a rational choice between two products, I have to think about my preferences; to make a decision, I have to somehow reason with regard to what the result of my decision will be if one of the many possible situations independent of my decision occurs (consider the decision as to whether to take an umbrella when it is raining or not raining outside). To make a choice in a strategic interaction, I have to think about what choices others would make based on their interests and the fact that they themselves are,

---

1. On the notions of deduction, induction and abduction see Chapter 2.

in all likelihood, reasoning about the choices I would make (a very simple way to grasp this point is to consider the Prisoner Dilemma, one of the most famous and debated examples in game theory).

As for the relationship between logic and reasoning, it is easy to see that logic deals with the rational connections between the sentences that we employ in our reasonings. Reasoning is an act of our mind, something *we do*. The rational and logical connections between sentences, however, remain ideally such even if we do not think of them, since they are not something that depends on us. So, in a sense, even logic deals with reasoning only indirectly. It is also true, however, that those connections between sentences have a role mainly in our reasoning activity, and therefore the position of those who consider logic as the normative theory of reasoning is also defensible.

All six normative disciplines which we have mentioned set standards of correctness against the background of which we can evaluate the goodness of our reasoning in cases of choice, decision, strategic interaction, estimation of a probability, solution of a logical problem. They do this by assuming certain standards of *rationality*, and by implicitly creating *idealizations* that may diverge from real cognitive agents, from our real cognitive and computational abilities, and more generally from the way in which we actually make decisions, choose between alternatives, and face calculations.

For example, logic sets *consistency* as a standard of rationality (absence of contradictory assumptions in a theory or set of sentences), and, in the case of deductive logic, distinguishes between valid and invalid inferences. A logically perfect agent, or a perfectly rational one from the logical point of view, is an agent which never falls into contradictory assumptions and is able to distinguish valid inferences from invalid ones without error. We can call *homo logicus* this "ideal" agent. It is clear that none of us complies with the *homo logicus* idealization in the sense just defined. Sometimes we find ourselves making contradictory assumptions, because their contradictory nature is not obvious – it requires complex reasoning to be identified, and the complexity of such arguments can give us problems. Furthermore, there are errors of logical reasoning to which we seem particularly exposed, so much so that they have been systematized in the history of thought and are usually called *fallacies* (we will look at some of them starting from Chapter 3).

Neoclassical economic theory and classical rational choice theory set a package of properties – mostly formal properties – that the agent's preferences must satisfy as the standard of economic rationality. For example, an agent's preferences are assumed to be complete and transitive. Completeness implies that any good is comparable with other goods, so that for any pair of goods x and y, either I prefer x to y, or y to x, or I think they are equally good. Transitivity dictates internal consistency of preferences – if an agent prefers x to y and prefers y to z, then they prefer x to z. Furthermore, some stability of preferences over time is assumed.

A perfectly rational agent has preferences that satisfy these properties (among others). We call this ideal agent *homo œconomicus*. Again, the misalignment with real agents is clear. For example, our preferences often change, yet this does not seem to imply that we are irrational (contrary to the standard of economic rationality above). Moreover, our preferences do not seem to be complete. Take an agent who has no interest in comics and ice cream. On what basis could these two goods be ranked in terms of "better", "worse", or "equal"? Even transitivity is not to be taken for granted.[2]

If we sometimes reason on the basis of contradictory assumptions, how can we justify the *consistency* assumption that we have seen above? If our preferences do not work as neoclassical economic theory or classical rational choice theory tells us, how do these disciplines justify the formal properties they impose in providing a model of choice-making? More generally, how can we justify the elaboration of reasoning models that seem to represent ideal agents (perfectly rational from a logical or practical point of view) but not actual agents, like ourselves?

One possible answer is that any theory that studies complex phenomena starts with idealizations. We cannot consider every aspect of our choices or strategic interactions. We cannot consider all the cognitive and computational limits that condition our reasoning or our calculations. We have to start from the essential aspects of these activities (those that make them work, in a sense) and abstract from the rest. In other words, neoclassical rationality is a *useful assumption* and, consequently, so are the properties that neoclassical economics and classical rational choice theory impose on preferences. Likewise, logical rationality – the assumption of consistency and infallibility in reasoning – is a useful assumption. That is, these assumptions are a useful construct for developing a series of simple idealizations and theories, with the implicit assumption that real agents are something else.

This answer is not immune to objections. Here is one of the many: an idealization is useful if it does not differ too much from the way things are; if it deviates too much, it no longer tells us anything illuminating. More specifically, if theoretical constructs are used to formulate explanations or predictions of real phenomena – for instance, to explain and predict how economic agents behave – we might as well start out from sufficiently realistic premises regarding the cognitive agents involved in choices, strategies, decisions, and reasoning.

Another possible answer is this: if we limit ourselves to *describing* how we actually reason, make decisions, interact with other decision-makers, choose goods, then we will not be able to say what is correct, or *rational*, and what is not. When we make choices, we make decisions, we adopt strategies, we pursue *our* goals. At the same time, we are aware of the fact that some decisions, choices, strategies that

---

2. An elementary illustration of this is the example of the *polite dinner party guest* in Amartya K. Sen, *Internal consistency of choice*, "Econometrica", 61, 1993, 495-521.

we happen to adopt are adequate to achieve these objectives (or more adequate than others), while others are not. However, if all that mattered were to describe how we make choices, we would not be able to distinguish proper choices from inadequate ones. If we did not distinguish between good and bad arguments (as we will do starting from Chapter 2), we would have no basis for objecting to an argument that does not seem to work. All we could do would be to report the fact that someone argues in a certain way.

The moral of what has just been observed is the following: We need normative approaches because we need normative standards for our reasoning. Compared to the previous answer, this one seems open to the idea that we need *both* approaches (normative and descriptive). Their combination allows us, at least ideally, to elaborate theories that do not lose sight of the role of normative standards, but at the same time start out from realistic assumptions about the agents who choose, decide, adopt strategies, and reason.

We will return to this point at the end of this chapter, in § 1.3, after having looked at some cases in which we do not reason as we should, and why. In particular, § 1.2 will focus on three *cognitive biases* concerning our reasoning in the context of decision-making, probability assessment, and estimates on a given question. Starting from Chapter 3, then, we will learn how to identify some important *logical fallacies*, or patterns of reasoning which may not be good but which, for some reason, we tend to follow. These do not have to do specifically with decisions or calculations, but with our reasoning in a great variety of different domains.

1.2 Heuristics and biases

We are well aware that we often do not think perfectly. Sometimes we get it wrong when we do mathematical calculations, or when we calculate probabilities. We contradict ourselves, we fall into logical fallacies, and so on. These errors are due to our cognitive and computational limitations and, as such, we are used to considering them as a simple "deviation" from correct calculation or reasoning procedures.

Some violations of the normative standards of reasoning, however, are not due *entirely* to our cognitive limitations, and, above all, they happen *systematically* in some contexts in which we make decisions, estimate probabilities, do calculations, or answer questions. Since the 1970s, cognitive psychology and decision psychology have focused on these phenomena. Researchers working in these disciplines have conducted experiments that have helped us not only see *how* concretely (and systematically) we deviate from standards, but have also proposed an *explanation* of why these deviations occur – what exactly makes them happen.

Among the most significant figures in these research areas we must undoubtedly include Daniel Kahneman and Amos Tversky, who conducted experiments on how we actually reason when we face decision-making problems, and when we answer questions that have to do with probabilities.[3] Their results gave a strong impetus to the emergence of disciplines such as *behavioral economics* and *behavioral decision theory*. The former focuses on the processes of economic choice of real agents (like "us") in the real world; the latter focuses more generally on the decision-making criteria that we actually put into practice. Both are based in part on the awareness that a series of errors and distortions systematically characterize these processes, and this awareness was made possible precisely by the experimental results and theories developed in decision psychology and cognitive psychology.

One general conclusion which emerged from experiments conducted over time is that, for a variety of reasons, when we have to solve a decision or calculation problem, we do not follow the logical-rational steps of the procedures that allow us to solve the problem (or at least, we do not follow them all), but we rely on *heuristics*, that is a series of "mental shortcuts" that allow us to tackle the problem *without* following the rational procedure to do so but, at the same time, in a way that is efficient in the majority of the cases we face every day. However, it has also been seen that this heuristic approach produces systematic biases, whose negative consequences are evident in specific contexts.

Many experiments by Kahneman and Tversky have contributed to the elaboration of the theory of heuristics and biases. In particular, they have shown the main types of bias that characterize our decisions, and proposed an explanation of their impact, thus opening the way also for methods to correct them. The theory of heuristics and biases has had great weight in behavioral economics, which in general uses many tools of empirical psychology, studying the processes that lead individuals to make certain decisions under given experimental conditions.

In a celebrated volume entitled *Thinking, Fast and Slow*,[4] Kahneman summarized many years of empirical and theoretical research in a general vision, and more precisely in the idea that our mind is made up of two cognitive systems, present to a greater or lesser extent in concrete individuals: System 1 and System 2. The first works in a fast, automatic, intuitive way, and is conditioned by our emotional states. The second, on the other hand, works in a slow, thoughtful way, allowing us to carry out complex calculations, to compare alternative choices, to follow plans. If System 1 proves effective in many situations, making our responses and decisions faster, in other situations, however, it deceives us, leading us to systematically make mistakes.

Like behavioral economics, the cognitive psychology of reasoning has tried to understand the reasons for some systematic reasoning errors, some typical difficulties and limitations of our inferential capacities. Models have been built that repre-

---

3. One of their most famous contributions in this perspective is Daniel Kahneman, Amos Tversky, *Prospect theory: An analysis of decision under risk*, "Econometrica", 47(2), 1979, 263-291.
4. Daniel Kahneman, *Thinking, Fast and Slow*, Farrar, Straus and Giroux, New York 2011.

sent and explain these aspects, indicating, when possible, some strategies to overcome the difficulties typically encountered by real reasoners.

In what follows, we will present three of the most famous cognitive biases. The *framing effect* and the *anchoring effect* emerged in the experiments of Kahneman and Tversky. The *hindsight bias* emerged in the experiments of two other psychologists of reasoning, Baruch Fischhoff and Ruth Beyth.

1.2.1 Framing effect

The expression "framing effect" refers to the phenomenon whereby if one and the same decision problem is presented in different ways, this will lead us to make different choices, even if the different ways of presenting it do not change the substance of the problem. To get a concrete idea, let us take a look at the experiment conducted by Kahneman and Tversky and known as the "Asian disease problem", named after the fictitious disease mentioned in the experiment.

In the experiment, two distinct groups of people are subjected to the same fictional story. At the end of the story, the first group is presented with a decision problem whose alternatives are A1 and B1, while the second group is presented with a decision problem whose alternatives are A2 and B2.

Group 1. [*Story*] A community of 600 people is affected by an Asian disease. If no sanitary measures are taken, the terrible pathology will wipe out the entire community. If instead we take health care measures, then [*Decision Problem*] we can choose one of two alternative programs:

- (A1) A program that saves 200 lives;
- (B1) A program that has a 1/3 probability of saving all 600 lives and 2/3 of not saving anyone.

Which program do you choose, A1 or B1?

In the experiment conducted by Kahneman and Tversky, 72% of the first group of respondents chose A1, and 28% chose B1. This trend was confirmed in further repetitions of the experiment.

Group 2. [*Story*: the same] [*Decision problem*] We can choose one of two alternative programs:

- (A2) A program that results in the death of 400 people;
- (B2) A program that has the probability of 1/3 that no one will die and 2/3 that everyone will die.

In the experiment conducted by Kahneman and Tversky, 22% of the second group of respondents chose A2, and 78% chose B2. This trend was confirmed in further repetitions of the experiment.

The crucial point of the experiment is that the A1 and A2 options are actually the same option, except that A1 is presented in terms of *gains,* while A2 is presented in terms of *losses.* The same goes for options B1 and B2. So, in reality, the two groups have been subjected *to the same decision problem*, but the problem is presented (*framed*) to the two groups in two different ways.

This diversity of wording is accompanied by a fact: where a choice is presented in terms of gains, as to group 1, agents are (in overwhelming percentages) *risk averse*, while where the same choice is presented in terms of losses, as to group 2, agents are (in overwhelming percentages) *risk taking*.

The facts emerging from the Asian disease experiment show us that we are not perfectly rational agents. Let us start with the fact, revealed by the experiment, that people make their decisions based on two different decision criteria: some follow a risk-averse criterion, and others a risk-prone criterion. The sample of the experiment is chosen in an adequately *randomized* manner by Kahneman and Tversky, and then by those who repeated the experiment. This means that, in terms of percentages, we should expect similar distributions of risk-averse people and risk-prone people. If we were perfectly rational, we would have to react to A1 the same way we would react to A2, and vice versa, because they are actually the same option. The same goes for B1 and B2. Consequently, in the two groups we should have very similar distributions for A1 and A2, which encounter risk aversion, and for B1 and B2, which encounter risk taking. Instead we have a real reversal of trends.

This is because, in cases like the one we are discussing, we are inclined to be sidetracked by the way the alternatives of a decision problem are presented, rather than rationally considering their actual consequences. Ultimately, Kahneman and Tversky's experiment contradicts the idea that we are perfectly rational decision-makers.

Kahneman and Tversky's experiment has even more drastic effects for a particular decision theory, which has long been dominant in decision theory and rational choice theory, namely the *Expected Utility Theory* (EUT), whose basic commandment is: Choose the action with the highest EU (expected utility).

Let us illustrate that with a decision problem. I have a sum of 5,000 euros available, which I can choose to commit to a given investment. Market signals say there is a very high probability that the investment will be successful. If this is the case, I will earn 15,000 euros by investing, but if it goes badly, I will lose half of the initial 5,000. If I don't invest, instead, I will simply keep my 5,000. I therefore have two possible *actions* (investing and not investing), two relevant *states of affairs* (good and bad market performance), and four possible *outcomes*, each of which gives me a specific utility (in this case, monetary utility):

Table 1.1 Investment example

	Good market trend	Bad market trend
Invest	15,000	2,500
Do not invest	5,000	5,000

Let us also assume that we attribute a probability of 70% that the market will perform well, and 30% that it will perform badly. Which of the two actions should we choose? EUT says to proceed like this: take an outcome of the first action (invest) and multiply its utility (15,000) by its probability – so we will have, in our example, 15,000 x 7/10 = 10,500; then do the same with the other outcome – we will then have 2,500 x 3/10 = 750. Add the two results – so we have 11,250. Proceed in the same way with the other choice (do not invest). The result of the sum is: (5,000 x 7/10) + (5,000 x 3/10) = 5,000. Choose the action that guarantees you the greatest expected utility: in this case, the action of investing.

EUT has been criticized for various reasons, although it still has a significant standing in several disciplines, including economics and some of its applications (e.g. public policy and health policy). Its adequacy as a normative theory of decision has also been questioned, but here we will consider only its *descriptive* or *empirical inadequacy* – that is, its inability to describe how, in fact, we real decision-makers choose between various alternative actions. Kahneman and Tversky's experiment was conducted precisely to show this inability on our part.

Let us now try applying EUT and see what choice it would prescribe in the Asian disease problem. For the sake of simplicity, we will give a value of 1 to a saved life, and 0 to a death from illness. Consider A1 and B1. The possible outcomes are: the salvation of 200 people and the death of the others (with probability 1) for A1, and the salvation of all 600 (with probability 1/3) or the death of all (with probability 2/3) for B1. The EU of A1 is therefore 200 (i.e. 200 × 1 + 400 × 0). What is the EU of B1? Exactly the same: 600 × 1/3 + 0 × 2/3, i.e. 200. The same calculation applies to A2 and B2, respectively, since they are different ways of presenting A1 and B1. Thus, for EUT, *all four alternatives* presented in the experiment are equivalent. Yet the experiment reveals a notable disproportion between the preferences for A1 and those for B1 (or those for B2 and those for A2). EUT is *descriptively inadequate*.

The framing effect plays a very important role in many concrete contexts of our everyday life. The following example shows its relevance to *consumer behavior*.

Participants in an experiment[5] were presented with meat and asked to rate its quality. The participants were divided into two groups. The product was presented to the two groups with different labels.

---

5. The description of the experiment and its results are in Irwin Levin, Gary Gaeth, *How consumers are affected by the framing of attribute information before and after consuming the product*, "Journal of Consumer Research", 15(3), 1988, 374-378.

Group A: The label indicates that the product has 75% lean mass.
Group B: The label indicates that the product has 25% fat mass.

The meat was however of the same quality for the two groups. In turn, the two groups were divided into two subgroups.

Group A1 and B1: had to evaluate the meat only from the label, which in addition to the percentage of lean/fat mass summarizes other characteristics, which are identical for both groups.
Group A2 and B2: tasted the meat, and did not just read the label.

The meat had to be evaluated by the two groups according to four different scales:

1. good taste *vs* bad taste;
2. oily *vs* non-oily;
3. high quality *vs* low quality;
4. fat *vs* lean.

For each of these dimensions a score from 0 to 6 could be assigned. Figure 1.1 shows the results of the experiment.

As you can see from the graph, meat whose label read "75% lean mass" received significantly better scores than that whose label reported "25% fat mass", both from the respondents in groups 1 (label only) and from those in groups 2 (label and

Figure 1.1 Meat evaluation

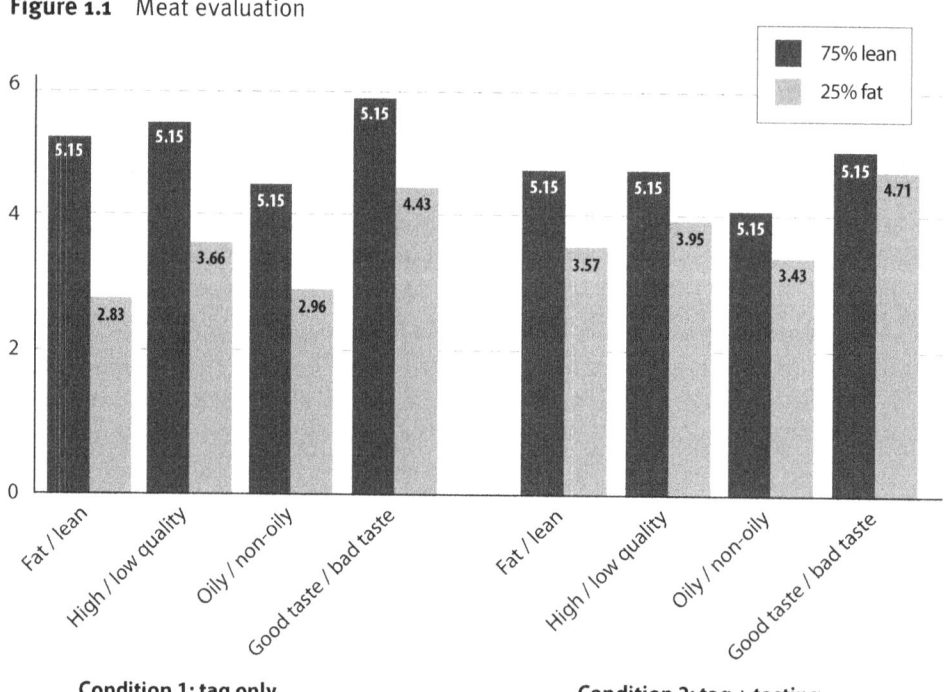

taste) – even if in groups 2 the disproportion in scores is less accentuated. In short, group A and group B evaluated *exactly* the same product in a different way, and this (i) regardless of the possibility of tasting or not, and (ii) in relation to the indication of lean and fat mass.

Point (ii) is crucial here: the two labels indicated the same percentages of lean and fat mass, but *presented them* in different ways. In short, not only did groups A and B evaluate the same product: they also found *the same information* on the label. However, they reacted not according to the meaning of this information, but rather to the way in which it was presented.

The experiments we have exemplified show that the way in which problems are framed influences the responses of the subjects in a decisive way, even if it does not tell us anything substantive about what is at stake in the problems themselves.

The cases in which the framing effect seems to have more marked effects on decisions are those in which the presentation of the problem triggers strong emotions in decision-makers, such as fear, panic or, conversely, trust. These emotions play a role in our decisions, not least because they block our rational considerations. In a situation where I have to make a decision about the lives of 600 people, being told that with the chosen program I will save 200 has a very different emotional impact from being told that I will have 400 die.

Emotions can also be called into question indirectly, leveraging on prejudices or on our cognitive or computational limitations. If I am given very strict time limits in choosing the health programs in the example, then my emotional involvement will have an even greater weight, because I will not have time to analyze the problem well, and to realize that (at least in terms of EU) the programs are equivalent.

In other words, in situations like these we tend to opt for what Kahneman calls "System 1" rather than what he calls "System 2".

Experiments such as the ones we have exemplified have a clear consequence: if we want to *predict* how an agent will make a decision, given some definite alternatives, we cannot reason about how the agent *should* make a decision, because normative theories – at least the classic, simpler ones – often prove inadequate in describing how agents choose. If our goal is to build economic models that explain and predict the choices of agents in given situations, we must take into account the fact that agents like ourselves are systematically influenced by biases, cognitive constraints and emotions.

The intertwining of economics and psychology is fundamental for this goal, because it allows us to consider the psychological processes and actual decisions of real economic agents, rather than the "rational" decisions postulated by neoclassical economic theory. Behavioral economic analysis was created precisely to meet this challenge and account for our decision-making activity as it really occurs, thus bringing economic theory closer to the real world. Furthermore, awareness of how we actually reason helps us notice our systematic biases, putting us in a better position to think critically, identifying errors and distortions, and avoiding them in our judgments and decisions.

Understanding how and to what extent biases affect our reasoning can also be important to improve some aspects of our practical interactions with other agents. For example, knowing about the results of the "lean/fat" experiment, a food business can take the opportunity to report that a given product has, say, 75% lean mass, rather than 25% fat mass.

1.2.2 Hindsight bias

The hindsight bias consists in overestimating the initial probability that an event would happen *after* that specific event has happened. To understand how it works, let us consider an experiment on the psychology of reasoning carried out by Baruch Fischhoff and Ruth Beyth[6] (here we modify it by reducing the possible answers from four to two). A *randomly* selected group of people is told the following story:

Nepalese War. Between 1814 and 1816 a war was fought in Nepal between the British and the Nepalese Gurkhas. The former enjoyed a better war organization and better armaments, the latter had a better knowledge of the territory and a stronger motivation.

How the war ended is not revealed, and people in the group are asked:

Based on what you have been told, what was *the likelihood of an English victory? And what* was *the likelihood of a victory for the Gurkhas?*

Taking the average of the likelihoods given by each individual in the group, Fischhoff and Beyth noted that the group attributed a likelihood of 67% to the British victory, and 33% to the victory of the Gurkhas. Then Fischhoff and Beyth divided the group into two subgroups. Subgroup 1 was told that the war had been won by the British, Subgroup 2 that the war had been won by the Gurkhas. Fischhoff and Beyth then repeated the initial question to each of the two subgroups.

It turned out that in Subgroup 1, the average likelihood attributed to a British victory had risen to nearly 90%, and that of a Gurkha victory dropped to around 10%. In contrast, in Subgroup 2, the average probability attributed to the victory of the Gurkhas rose to around 65%, and that of the British fell to around 35%.

What is wrong with the last two answers? The experiment asks to evaluate the initial or *a priori* probability of the British victory and the complementary one of the

---

6. Baruch Fischhoff, Ruth Beyth, *I knew it would happen: Remembered probabilities of once-future things*, "Organizational Behavior and Human Performance", 13, 1975, 1-16.

victory of the Gurkhas. Just to have a clear idea of what this is, the *a priori* probability that a 2 comes up from the roll of a (non-loaded) die is 1/6. More generally, the *a priori* probability of an event E (e.g. "A 2 comes up", "The Gurkhas win the war", etc.) is the probability we attribute to E *before* the event occurs (before we roll the given die, before the war begins) – or, more precisely, before we are given *additional information* about the event. The *a posteriori* probability of E, on the other hand, is the probability that we attribute to E *after* (we come to know that) it occurred – or, more precisely, after we are given additional information about it. What is the *a posteriori* probability of the "A 2 comes up" event? Simple: if the roll gave 2 as a result, the *a posteriori* probability of "A 2 comes up" is 1; otherwise, it is 0. This however *cannot modify* the *a priori* probability of "A 2 comes up", which always remains 1/6 (if the die is not loaded).[7]

Now let us understand what is wrong with the subgroups' answers: the fact that the average attributed probabilities are different from those of the initial group means that at least some individuals have *changed* their estimate of the *a priori* probability of the British (or Nepalese) victory. But this is just as incorrect as thinking that "A 2 comes up" had an *a priori* probability greater than 1/6, once we know that a 2 actually came up. The information on who actually won the war is irrelevant for evaluating the *a priori* probabilities of "The British win" and "The Gurkhas win", just as the information on which number came out is irrelevant to assess the *a priori* probability of "A 2 comes up".

The occurrence of the hindsight bias demonstrates that we are not perfect probabilistic reasoners. In particular, we tend to reason incorrectly about *a priori* probabilities when we are given information about how things really went. The effect is explained by the psychological difficulty of ignoring the information received, which remains difficult to disregard even when we are explicitly asked not to consider it.

The study of the hindsight bias has also revealed a phenomenon that accompanies the overestimation of the probability of events actually occurring in the past: agents tend to rationalize their responses by giving, *a posteriori*, greater importance to the aspects of the problem that suggest an answer in line with the outcome they have been informed of, and giving less importance to those that suggest answers that are not in line with it. In the case of the Nepalese war, people in Subgroup 1 (who were told the British won) tend to rationalize their overestimation of the *a priori* probability of "The British win" by insisting on the superior military organization of the British, the discipline of troops, the greater availability of weapons, etc. People in Subgroup 2, on the other hand, rationalize their response by insisting on the best knowledge of the territory, the stronger motivation, the effectiveness of guerrilla tactics, etc.

---

7. Probabilistic reasoning will be the topic of Chapter 10.

Essentially, hindsight operates by leading us to rationalize the overestimation of the probability, as if the event that actually occurred would necessarily be more predictable than the alternative events that did not occur. In fact, in this rationalization we end up believing that the event that happened was more predictable than it really was. (If you are curious to know how it really ended, the war was essentially won by the British, even if the victory involved a peace treaty and mutual concessions.)

The hindsight bias plays a role in many contexts, including some that are very familiar to us. Take, for example, the evaluations of sports competitions – like a football match. When we evaluate the outcome of a match retrospectively, we are all ready to say that the outcome was predictable and to explain it: those who lost should not have played one way but another, some player substitutions should not have been made, or other players should have entered the field from the bench, etc. In short: we try to explain not only why the match ended in a given way, but also why the factors for that outcome weighed more than the opposing ones. Obviously this approach is not always justified: in almost all sports (especially team sports) it can happen that some matches are decided by more or less random episodes, which have little to do with the overall performance of the teams on the pitch.

Similarly for electoral competitions: before the elections we attribute a degree of uncertainty to events ("Who will win?") which we then reduce in our *a posteriori* judgments. An exemplary case is provided by the 2016 US presidential election: many observers, American and international, had considered Donald Trump's victory as highly unlikely, especially due to the assumption (whether well founded or not) that what Trump offered matched no substantial need of the electorate. After the tycoon's victory, some observers reviewed their assessment, coming to the conclusion that Trump had managed to leverage (perhaps only for propaganda purposes) the needs of a specific socio-economic segment of the electorate. Belief revision can be the right thing to do when new information is given, but, as the die example above shows, it is wrong to revise the *a priori* probabilities given the *a posteriori* ones.

The hindsight bias also has important practical consequences, and awareness of it has led to the need to stem it in some areas. For example, some legal systems require that information of a certain type should not be provided in court when certain types of disputes have to be decided. This is because, by receiving that information, whoever has to decide the dispute would overestimate the *a priori* probabilities that the event in question would happen. Let us consider the following example.

A shopper is walking around a supermarket looking for some products. However, he does not notice a banana peel on the floor, he slips on it, and breaks his leg. After the accident, he learns that the management of the supermarket has decided to improve cleaning by increasing the number of shifts. Before the accident the attendants cleaned the floors every four hours, now they clean them every two hours. The customer sues the supermarket claiming that the accident was due to negligence in ensuring the cleanliness and safety of the place, and argues that the new, more frequent shifts have been implemented precisely to remedy this negligence – that is, because the previous shifts were not sufficient to guarantee an adequately low risk of falling.

What happens in the example is clear: the customer (even unwittingly) leverages the hindsight bias of the decision-makers. Given the accident and knowing that cleaning shifts have been increased, he argues as if the accident was more predictable and avoidable than perhaps it was. What the customer argument "sells" to the decision-makers is this: "The fact that cleaning shifts have been increased shows that more could and should have been done".

Some legal systems, for instance US federal law, establish that such information cannot be admitted as evidence in a trial (Federal Rule of Evidence 407). In our case, the plaintiff (the customer) cannot tell the jury that the cleanliness and safety of the supermarket was improved after the accident. This information cannot therefore be assessed by the jury in making a decision on negligence and on the right to compensation. The reason is that the jurors' task is to assess the negligence on the basis of the *a priori* predictability of the accident. The hindsight bias shows that information on what then happened would compromise that assessment.

An interesting point is that other legal systems, for example the Italian one, allow this kind of information as evidence that can be used in a civil case in which compensation after an accident is claimed. This is despite the effects of the hindsight bias. Why are the systems different in this respect? Consider that, in a system like the Italian one, it is judges with specific legal skills who decide cases like that, not a jury made up of ordinary people and without specific competences. There is an assumption implicit in the idea that a judge can evaluate information regarding what happened better than a jury; the assumption seems to be that the professional education and experience of judges shield them more from possible prejudices, errors, fallacies and biases. Is this assumption justified? Can it be empirically and experimentally shown that professional decision-makers are less affected by biases?

We happen to apply this assumption to other professional skills as well. For example, we tend to believe that professional investors who make economic or financial decisions are more capable of resisting biases than others.

However, some studies show that the economic signals as to the prospects of a certain company are valued differently by professionals who are told that the company then went bankrupt compared to those who are told that the company did well. So, in fact, having relevant professional skills is not enough to insure us against hindsight bias: even competent people can fall victim to it.

1.2.3 Anchoring effect

In many situations, we find ourselves having to make quantitative estimates or calculations based on an initially given value, which we adjust in order to provide our estimate or the result of the calculation. It is not uncommon, however, that, in making our estimates or calculating the result, we allow ourselves to be influenced by the initial value in an incorrect way, even when it is actually irrelevant, and we are thus led to provide estimates or results that are not correct or really justified.

The anchoring effect consists in the disposition of cognitive agents to "anchor themselves" to data or values initially provided in tackling a problem, in a way that does not actually guarantee a correct calculation of the final value, or a rational consideration of the problem.

Let us look at some examples of how the anchoring effect works in practice. In one of the first experiments that revealed the incidence of this effect, Kahneman and Tversky divided the participants into two groups, who were asked, respectively:

Group 1: estimate in 5 seconds the result of $1 \times 2 \times 3 \times 4 \times 5 \times 6 \times 7 \times 8$.
Group 2: estimate in 5 seconds the result of $8 \times 7 \times 6 \times 5 \times 4 \times 3 \times 2 \times 1$.

The figures given in the two products are exactly the same, and since multiplication is commutative, the result is the same (i.e., 40,320). Of course, five seconds is not enough for the vast majority of us to calculate the two products correctly, and Kahneman and Tversky did not expect to receive (many) correct answers. Almost all the answers were indeed wrong, but it is remarkable how the errors were distributed in the two groups: in Group 1 (with the ascending series) the median estimate for the result of the operation was 512, in Group 2 (with the descending series) it was 2,250. This means that most of Group 1 gave much lower estimates not only of the correct answer, but also lower than the results given by most of the members of Group 2.

If the operation was the same, what explains this stark divergence between the two groups? The explanation of the difference lies in the psychological anchoring occasioned by the first numbers of the series. Let us see why, and how.

The only difference in the questions for the two groups is in the order of the factors, which is not relevant in the multiplication. However, the order of the factors led Group 1 to calculate starting from the lowest figures (1 x 2 x 3 x ...), and Group 2 starting from the highest (8 x 7 x 6 x ...). Given the time limitation, the members of the two groups did not perform the entire calculation: at some point they stopped and "adjusted" the result they had actually calculated, increasing it a bit. For example, someone in Group 2 calculated 8 x 7 x 6 x 5 = 1,680, and then intuitively adjusted the result. In other words, that individual anchored herself to the calculated result, and adjusted it. But adjusting when starting from 8 x 7 x 6 x 5 = 1,680 is very different from adjusting when starting from 1 x 2 x 3 x 4 = 24, because the "anchors" in the two cases are very different.

In short, in the few seconds available, the participants perform part of the multiplications and then make an intuitive adjustment, which is on average insufficient.

When the data considered are reliable indicators of the overall data, anchoring can work as a good heuristic. Ideally, if I was able to calculate 8 x 7 x 6 x 5 x 4 x 3 = 20,160, then I may well adjust by making an estimate close enough to 40,320, since I am starting from an "anchor" value (20,160) which is not very far from the result of the 8 x 7 x 6 x 5 x 4 x 3 x 2 x 1 operation. If the data considered are too far from the overall data (or more generally, if they are not reliable indicators), then anchoring leads to significant error.

There are, however, partially different explanations for the hold that anchoring has on us. For some it is a form of mental laziness; for others it arises from the reasonable inclination to use available data; for still others it connects to further psychological mechanisms including suggestion.

The occurrence of the anchoring effect can also be seen in situations where calculations are not required but simple estimates are. For example, in one study, two groups of students were asked what the average price of a textbook in the campus bookstore was. Members of one of the two groups were also asked whether, in their opinion, the average price was higher or lower than $ 7,163.50. Those who were initially exposed to this absurd figure gave on average higher responses than those of the other group. Once an initial figure was presented, it acted as an "anchor", even if it was absurd and, in fact, irrelevant to the estimation that the participants had to make.

The anchoring effect also has a very important impact on our activities. Let us go back to the example proposed above: a customer breaks his leg by slipping on a banana peel in a supermarket and takes legal action against the supermarket management, claiming that the latter had been negligent in ensuring proper cleaning of the premises. Suppose the customer is able to convince the decision-makers of the supermarket's negligence, even without using the information regarding the increase in the frequency of cleaning shifts. Now the customer must quantify the harm suffered, hoping that he will be granted this sum or a little less. Let us suppose that he asks for the considerable sum of 655,000 euros, claiming that in addition to the medical expenses incurred (quantified as 25,000 euros paid to a specialized clinic), he suffered from a loss of earnings, as he could not participate in an important work activity that presumably would have earned him an estimated sum of 630,000 euros.

Suppose in addition that the judge is not entirely convinced, as he suspects that the plaintiff's assessment is exaggerated. Let us also assume, however, that the judge awards to the plaintiff a not much lower sum, for instance 610,000 euros. It cannot be ruled out that the judge was a victim of the anchoring effect. The request of the plaintiff was the quantitative estimate discussed in the case, and this may suggest that it acted as an "anchor" (i.e. an implicit reference point) in the judicial decision, without the judge being aware of it.

The experiments on the anchoring effect show that adjustments of this kind are made because the starting request acts as a psychological anchor, as a reference point, however exaggerated it may be in some cases. If it could be shown that in the case of the supermarket the judge would have decided not on 610,000 (as supposedly decided) but on 120,000 if the victim of the accident had made a claim for 150,000, then the effect in question could be confirmed.

An area in which the anchoring effect plays a decisive role is that of negotiations, or bargaining, especially when participants need to establish a price or a quantity. Experienced participants, in these cases, open the negotiation with a request higher than what they themselves find appropriate. This will lead the counterparty to make an offer not much lower than the initial request. This also happens because an overly marked divergence with respect to the previous proposal would lead to the end of the negotiation. It must also be considered that in many such negotiation or bargaining contexts, not all parties have a sufficiently defined idea of the value of the asset they are bargaining about. The ideas we have regarding how much an asset is worth can be quite vague, and in these cases the anchoring effect becomes even more pronounced. Once the initial proposal for an asset has been received, the counterparty will adjust its rather indeterminate estimate by taking the proposed value as a reference point, without deviating too much from it. For example, suppose someone proposes a price of 100,000 euros for an apartment that, in the estimation of the proposer, is worth 75,000. The proposer believes that the possible buyer particularly wants that good and is willing to pay up to 90,000 euros. If the buyer does not have a definite idea of the value of the asset and is subject to the anchoring effect, she will make her first counter-offer not deviating too much from the value of 100,000 euros.

These strategic interactions also require a certain psychological sensitivity, because the negotiator must be able to guess the potential availability of the counterparty and must not start from an excessive overestimation, given the risk of losing credibility and compromising the negotiation.

1.3 The value of normative standards

The three biases we have seen give us concrete examples of how in certain cases we do not respect the normative standards of rationality required from our reasoning. Hence, what to do with the normative dimension of reasoning, given that we happen to systematically deviate from it? In § 1.1 we saw, somewhat abstractly, that we still need normative standards – otherwise we would not be able to say that a choice, or a decision, or a strategy is adequate (or inadequate), and we would not have a starting point to contest an argument (or agree with it): these aspects are foreign to the pure *description* of how we actually reason.

In addition to this, there is one thing (somewhat less abstract) that we can learn from the three biases we have considered. Once we are told where the reasoning error is, *we* recognize it as such. Suppose we are subjected to the Nepalese war experiment, and we adjust the *a priori* probabilities based on the information we are later given about how the war actually went. Then suppose we are told what we did, and how *a priori* and *a posteriori* probabilities work. At that point we understand that we were wrong, in the same way as we understand that it is wrong to think that the probability of "A 2 comes up" was 1, and not 1/6, just because after the roll, 2 is the number that actually came up. Suppose we choose program A1 in the Asian disease problem and then we are told that we should have chosen according to the Theory of Expected Utility. We would recognize that the two alternatives were actually equivalent. In other words, normative standards are not simply rules that for some reason some discipline seeks to impose: we ourselves recognize the value of normative standards (the rules of probability calculus, for example) and admit that in some cases we make mistakes.

This, obviously, is not just our impression: if the purpose of our decisions and strategic choices is to pursue our interests, then we need criteria that separate the decisions and choices suitable for that purpose from those that are not. But that means we need normative standards. If the ideal end of reasoning is the attainment of truth, we need something that tells us what inferences and logical operations must or can be performed for this purpose – that is, what forms of reasoning ensure that we infer true conclusions from true premises, or (highly) plausible conclusions from true premises. In other words, we need the normativity of logic.

In any case, the fact remains that a normative theory of reasoning can give us an illuminating perspective only if the standards it imposes do not presuppose cognitive agents who are *too* idealized, as we observed in §1.1. Following the experiments in decision and reasoning psychology, many researchers in the normative disciplines we have listed have insisted on the importance of incorporating empirical findings into decision, strategic choice, and rational choice theories. In short, they stressed the need to incorporate those results into idealized models of decisions and choices, and to combine the need to provide normative standards with the need to take as realistic a starting point as possible as regards the agents who make decisions and adopt strategies. This has led to the birth of some disciplines that have acquired more and more importance over time, for example:

- behavioral economics;
- behavioral decision theory;
- behavioral game theory.

Arguably, there are reasons why this trend has not extended to logic and probability theory – we do not have "hybrid" disciplines between logic and the psychology of reasoning, for example. Let us take a brief look at these disciplines.

Logic is concerned with patterns of reasoning that can lead us to true or at least probably true conclusions. Contrary to the disciplines that deal with choices, decisions, and strategies, logic is not meant to predict how an agent will try to arrive at true conclusions – in other words, its task appears to be purely normative. Consequently, even if the standards set by some logical theories can be changed to define an ideal logical agent that is more plausible, there has not been the same need to define a discipline that integrates the normative dimension of logic with the results of the psychology of reasoning. However, consider that disciplines such as Critical Thinking and Argumentation Theory try to place logic at an intermediate level, which preserves the normative perspective while at the same time making sure that it is truly usable by real cognitive agents – or, in other words, making standards closer to reality.

1.4 Conclusion

In this chapter we have introduced the distinction between the normative and descriptive dimensions in the study of reasoning. As a consequence, there is a distinction between the disciplines that focus on the first dimension (normative disciplines) and those that focus on the second (descriptive disciplines). We have seen how Critical Thinking fits into this distinction. We have discussed the fact that we real cognitive agents do not always comply with the normative standards of reasoning, and we asked what justifies the normative approaches (§ 1.1). By exploring some cognitive biases, we have seen that, in some contexts, we are systematically exposed to reasoning errors, and why. We have discussed the relevance of descriptive disciplines in this area, and referred to the theory of heuristics and biases, as well as the theory of "thinking fast and slow" (§ 1.2). Finally, we have seen why, although they do not describe how we reason, normative approaches remain indispensable (§ 1.3).

2 | What is an argument?

In this chapter we define what an argument is and introduce the important notion of argumentative structure (§ 2.2). We then distinguish between simple and complex arguments (§ 2.2.1). This body of knowledge will allow us to identify arguments – that is, to recognize which sequences of sentences are arguments, and which are not. However, these notions do not tell us how to evaluate an argument. To do this, we will introduce the notion of a good argument (§ 2.3), and exemplify some types of arguments and reasoning: deductively valid arguments and deductive reasoning (§ 2.3.1), and their relationship to good arguments (§ 2.3.2). Next, we will illustrate some forms of non-deductive reasoning (§ 2.3.3) such as inductive reasoning, abductive reasoning, and reasoning by analogy, and discuss an example involving inductive reasoning proposed by the British logician and philosopher Bertrand Russell. We will close the chapter by discussing the notion of the strength of an argument, which applies to non-deductive reasoning, as distinguished from the persuasive effectiveness of an argument (§ 2.3.4).

2.1 Introduction

If we think about it for a moment, we formulate arguments on countless occasions in everyday life: when we have a chat at the bar, participate in a discussion on social media, take a university exam, write a dissertation, and so on. In fact, in daily life it happens to each of us to make *assertions*, that is, to say how things are. If we are sincere in making an assertion, then we believe that things are such as we say they are. Precisely for this reason, we usually assume that others must agree with us in this respect. If things are such as we say they are, others must also believe that they are like this. However, especially in certain contexts, disagreement can arise, that is, it may happen that one person believes that things are in a certain way and another person believes that things are different. In these cases, it is not enough to make an assertion to obtain consensus. It is also necessary to say something in support of what is asserted.

If we want others to think as we do, that is, to believe that things are the way we think they are, we can try to convince them that things are just like that. Imagine, for example, that you want to convince someone that dolphins are mammals, that the truth must always be told, that global warming is caused by air pollution, that the Covid-19 epidemic will have disastrous consequences for the world economy, that if the U.K. had introduced the lockdown two weeks earlier the number of deaths caused by the virus would have been much lower. There are many ways to obtain the consent of others. It can be done by deception or by force. However, the most rational way to do this is to formulate an argument, i.e. to give people reasons why they should believe that what we say is true, or that one should take a certain action, that something has caused something else, that a certain event is likely to happen, that things would have gone thus-and-so if a certain thing had happened, and so on. Providing an argument for a certain sentence therefore means providing a series of considerations that *support* that sentence. It could be claimed that the ability to argue is a characteristic that sets us apart as human beings, a capacity by which we try to persuade others to agree with what we think.

Although all of this is ultimately obvious, it is less obvious what an argument is.

2.2 Arguments and argumentative structure

An argument is a sequence of sentences, called *premises*, which provide reasons in support of another sentence, called *conclusion*.[1] The role of the conclusion is to express something that we intend to support and on which we believe others should agree. The role of the premises is to express considerations intended to ensure that this agreement takes place, or to induce others to accept that the conclusion is true on the basis of the (supposed) truth of the premises and their rational connection with the conclusion. In other words, arguing in support of a conclusion is equivalent to providing a *justification* for the latter, that is, to set out the reasons we have (or at least think we have) for believing that it is true, with the expectation that that these reasons will be accepted by others.

However, it is not always easy to identify the premises and conclusions of an argument. If we are lucky, we find words in verbal or written communication that indicate that a certain sentence serves as a premise or as a conclusion, and we can use these words to "signal" to other people what the premises and conclusions of our argument are (or we think they are). Some of these indicators are listed in Table 2.1.

However, we cannot rely only on these indicators to identify the premises and conclusion of an argument, because often these words do not appear in the argumentation. To identify premises and conclusion we must learn to play a more abstract game: that of understanding which sentences in an argument provide reasons for some other part of the argument, and which sentence is supported, in the argument, by other sentences.

---

1. In some cases, an argument may include only one premise. In Chapter 4, we will exemplify a circular argument that includes only one premise. In general, however, an argument needs more than one premise to fulfil its function – that is, to provide reasons for the conclusion.

Table 2.1 Conclusion and premise indicators

Conclusion indicators	Premise indicators
So	Because
Hence	Since
Therefore	Assuming that
As a consequence	Given that
Thus	In fact
Consequently	As
For this reason	Why
It follows that	In that
etc.	etc.

The fact that a sentence actually provides reasons in support of another sentence depends on the *structure* of the argument. Taking up a metaphor proposed by Andrea Iacona,[2] the argumentative structure can be conceived as the skeleton that supports the sentences of which an argument is composed, so that the premises provide reasons for the conclusion. Just as the skeleton of the human body is not visible to the naked eye but can be described by looking at the movements of the other parts of the body, in the same way the argumentative structure can be described by observing how the discourse is articulated. Using another metaphor, we could say that the argumentative structure is the glue that connects the premises of an argument to its conclusion.

It may not be trivial to identify the argumentative structure of an argument. For example, the ideal way of presenting an argument follows a precise order (in which the premises precede the conclusion) and includes all the premises necessary to adequately support the conclusion, but in fact we do not always argue like this. It may happen that the conclusion is formulated before the premises, or that some premises are not made explicit and remain hidden, or that the conclusion is missing. Let us imagine that in the course of a discussion someone says:

> Those banks that embrace the Amazon model remain competitive in the market. Indeed, the Amazon model offers new purchase and management online capabilities, and the banks offering those capabilities remain competitive.

The discourse here opens with the conclusion of an argument, while the following sentences provide the reasons that support the conclusion. The argument can therefore be reformulated as follows:

---

2. In Andrea Iacona, *L'argomentazione*, Einaudi, Torino 2006[1], p. 31.

> Only banks that offer new online purchasing and management capabilities will remain competitive on the market; the Amazon model offers new purchase and management online capabilities; therefore, banks that embrace the Amazon model will remain competitive.

This reformulation follows the ideal order in which the premises come before the conclusion. Note that this makes it easier to see that the conclusion follows from the premises.

Now consider the next argument:

> The Amazon model provides new purchase and management online capabilities and only banks offering this kind of service remain competitive.

In this case, the argument appears incomplete: the conclusion is missing. Nonetheless, the conclusion – namely that only banks adopting the Amazon model will remain competitive on the market – is easily identifiable in the light of the premises that provide reasons to support it. Let us examine another example:

> Only those banks that offer new purchase and management online services remain competitive in the market. Therefore, a bank which embraces the Amazon model will be competitive.

Here a premise is missing, i.e. the premise according to which the Amazon model provides new online purchase and management capabilities.

In this volume, we will usually present or discuss arguments in their complete form. Each argument will include the conclusion, all the premises necessary to support it, and premises will come before the conclusion. In general, an argument in its complete form can be schematized as follows:

$$
\begin{array}{l}
\text{Premise 1} \\
\text{Premise 2} \\
\ldots \\
\underline{\text{Premise n}} \\
\text{Conclusion}
\end{array}
$$

From now on we will use this scheme throughout the volume, and we will use "P1" for "Premise 1" (and so for other premises) and "C" for "conclusion".

Now, one might wonder why arguments are not always presented following the ideal scheme outlined above, which makes them easier to understand. There are several pragmatic reasons for this. For example, if we do not explicitly state the conclusion of an easy argument, our interlocutors will be induced to take the last step by themselves, and this will prompt their attention and involvement in the discussion. However, if the argument proves to be more complex for the interlocutor than we thought, this advantage will be lost, because the interlocutor could be confused or unsure about the conclusion to be drawn. As for the inversion of the ideal order between premises and conclusion, sometimes we want our interlocutor to focus immediately on the conclusion, i.e. we want the "message" we intend to support to get through immediately, and, to that purpose, we put forward the conclusion before presenting the reasons that support it. This inversion diverges from the rational argumentative order of our claims, but this is a low price to pay if those we are addressing are well capable of understanding the structure of the argument we are proposing. Finally, sometimes we omit one or more premises not because they are not important or even necessary, but because we believe that our interlocutors are already assuming them as a part of their background beliefs or knowledge. The clarification of such premises is therefore not necessary for the purpose of communication, because our interlocutors will integrate the argument themselves. However, we can be wrong and attribute to our interlocutors' assumptions, knowledge or beliefs that they do not actually have. In this case, the omission of one or more premises will make the argument weaker, because the interlocutor may believe that the premise that needs to be added is not true, and therefore should not be assumed.

Distinguishing the premises from the conclusion, and identifying the argumentative structure that articulates their relationship, is very important in order to recognize the point under discussion, what type of argument we are confronted with, and to evaluate the validity of the argument. In fact, we cannot evaluate whether an argument is a good one (we will see in § 2.3 what this means exactly) if we have not first correctly identified the premises and the conclusion of the argument in question. In Chapter 4 we will also see that identifying the structure of the argument is important from a dialectical point of view too: it is a fundamental step in formulating good counter-arguments and in carrying out a refutation.

2.2.1 Simple and complex arguments

In the examples considered so far, we have presented and discussed *simple arguments*, i.e. arguments consisting of one or more premises which provide reasons to support a conclusion. Arguments, however, can take on a more complex structure, in which several simple arguments are linked together. The most common types of complex arguments are (1) those in which the conclusion of one argument (called "subordinate") is the premise of another argument (called "main argument"), and (2) those in which two or more independent arguments provide reasons for the same conclusion. In the latter case, different arguments are linked together in order to make the conclusion more solid and convincing. Here it is an example:

> All the service companies that have become one-stop shops capable of responding to the specific needs of the customer are successful today. In-

deed, these companies aggregate their products alongside innovations made available by third parties. Today, US bank services work as one-stop shops and are therefore successful.

Here we are faced with a complex argument of the first kind, consisting of a main argument and a subordinate one. The main argument can be schematized as follows:

(P1) Service companies that have become one-stop shops are successful today.
(P2) US bank services operate as one-stop shops.

(C) US bank services are successful today.

The subordinate argument provides reasons in support of premise P1 of the main argument, and can be outlined as follows:

(P1) Service companies that have become one-stop shops, which aggregate their products to the innovations made available by third parties, are able to respond to specific customer needs.
(P2) Companies that respond to customer needs are successful.

(C) Service companies that have become one-stop shops are successful.

Here the subordinate argument serves to give reasons in support of a premise of the main argument, which appears to be of particular importance to uphold the conclusion. On closer inspection, in turn, the P1 of the subordinate argument is itself a complex argument, which can be reconstructed as follows:

(P1) Companies operating as one-stop shops aggregate their products with innovations made available by third parties.
(P2) Companies that aggregate their products with innovations made available by third parties meet the specific needs of the customer.

(C) Companies operating as one-stop shops meet the specific needs of the customer.

Thus, in this example we actually have *three* linked arguments where the premises of each of the first two provide support for one of the premises of the next argument.

A complex argument of the second type is the following:

Online platforms make financial services more competitive by allowing them to better respond to the clients and also offer a higher level of transparency and comparability of financial products.

In this case, two independent arguments are used to support the same conclusion. The first one can be schematized as follows:

(P1) Anything that makes it possible to meet the client's specific needs makes financial services more competitive.
(P2) The new online platforms make it possible to respond to the client's specific needs.

(C) The new online platforms make financial services more competitive.

The second argument is structured as follows:

(P1) A higher level of transparency and comparability between financial products makes financial services more competitive.
(P2) The new online platforms guarantee a greater level of transparency and comparability of financial products.

(C) The new online platforms make financial services more competitive.

In this case, the two arguments actually support the same conclusion, giving different reasons in its favor. In general, complex arguments either strengthen a premise that may be controversial, or strengthen the conclusion, so as to more easily persuade the audience to accept it.

Note that complex arguments do not contradict our definition of an argument as a sequence of sentences that provide reasons for another sentence. In complex arguments of the second type, it is the case that some subsets of the premises taken individually are already sufficient to provide reasons for the conclusion. For example, saying that "Anything that makes it possible to meet the client's specific needs makes financial services more competitive" and that "The new online platforms make it possible to respond to client's specific needs" is sufficient to give reasons to believe that "The new online platforms make financial services more competitive". The same can be said of "A higher level of transparency and comparability between financial products makes financial services more competitive" and "The new online platforms guarantee a greater level of transparency and comparability of financial products".

In complex arguments of the first type, one or more sentences play a double role: they are the conclusion of the subordinate argument, and a premise of the main argument. But we still have sequences of sentences that provide reasons for something else, and sentences that are supported by others. This also makes it apparent that the game we are supposed to play when we are called upon to identify the parts of a complex argument is, fundamentally, the same as that we play for simple arguments. In both cases, we are to establish what provides reasons to support what. In the case of complex arguments, the game is just a little bit more complicated.

2.3 Good arguments vs. bad arguments

So far, we have defined what an argument is, we have then examined its constituent elements, and distinguished simple arguments from complex arguments. All this provides us with a method for identifying an argument, its parts, and how those parts fit together. However, it is not yet clear what distinguishes a good argument from a bad one. In other words, under what conditions do the premises of an argument provide *good* reasons to support its conclusion? To understand this, we need to consider some additional characteristics of arguments.

Let us go back to our daily experience for a moment. When we formulate an argument, or somebody else proposes an argument to us, two distinct claims are made: (1) that the premises are *true*, and that (2) they provide *reasons* to support the truth of the conclusion.

This observation, however elementary, allows us to give an adequate definition of "good argument".

> *Definition of a good argument*
> An argument is a *good* one if and only if:
> (1) All its premises are true.
> (2) The premises provide reasons to support the conclusion.

In the event that at least one of the conditions (1) and (2) is not satisfied, we are instead faced with a *bad* argument. To clarify this, we can consider a famous example. In 1926 the Romanian sculptor Constantin Brâncuși was invited to exhibit some of his works at the Brummer Gallery in New York. Among the pieces sent from Paris to New York was a sculpture titled *Bird in space* (Figure 2.1), now valued at over $ 26 million.

Upon arrival in New York, the crate containing the sculpture was inspected by the customs officer, who found a smooth, tapered bronze object inside, an object that looked to the officer like a kitchen utensil rather than a bird. In accordance with the law of the state of New York, the officer therefore applied the customs duty for bronze artifacts intended for sale, refusing to qualify *Bird in space* as a work of art, which was instead free from taxation according to the law. This enraged Brâncuși, whose recognized reputation as an artist was questioned by an obscure customs officer. Brâncuși then decided to resort to a lawyer and bring the matter before a court, which was asked to determine whether *Bird in space* was actually a sculpture, and therefore whether the customs duty was due or not. After a long and interesting trial, the judges agreed with Brâncuși and stated that *Bird in space* was indeed a sculpture. As a consequence, the court ordered the tax paid to be reimbursed.

Regardless of the outcome of the trial, we are interested in two points here. First, the lawyers of both sides proposed arguments in order to persuade the judges to recognize their claims. So, the success of an argument was a key element in an extremely practical issue: the payment of a sum of money, or the exemption from payment. Arguments, therefore, can prove to be crucial in very concrete aspects of our life. Second, the arguments presented by those lawyers have characteristics that make them very interesting – even if only one of them was considered a good argument. Let us examine the exchange of reasons that took place in courtroom more closely.

According to the customs office, *Bird in space* was not a sculpture. In *United States v. Olivotti*, a previous case decided in 1916, the court had established that an artifact is a sculpture only if it is "the imitation of natural objects". *Bird in space*, however, did not faithfully mimic the shapes of a bird. The figure was devoid of feathers and wings, the body appeared elongated and rounded, the head and beak were reduced to an inclined oval surface. Nobody would have said it was a bird. We were therefore not dealing with a sculpture but with an artifact of common use, whose importation was not duty-free.

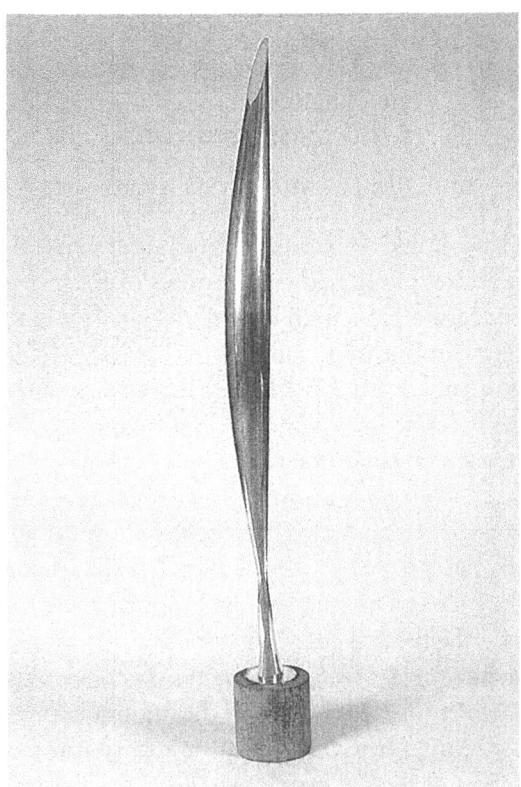

Figure 2.1 Constantin Brâncuși, *Bird in space*, 1926

Source: Art Poskanzer

The main argument of the customs lawyer (CL) can be schematized as follows:

(P1) All sculptures imitate natural objects.
(P2) *Bird in space* does not imitate a natural object.

(C) *Bird in space* is not a sculpture.

Brâncuși's lawyer argued instead that *Bird in space* was in fact a sculpture. The *Olivotti* case had been decided ten years earlier and, thus, did not constitute a relevant precedent. In that period of time, new artistic currents had developed which aimed to represent abstract ideas and not to imitate nature. In *Bird in space* the sculptor did not pay attention to the physical characteristics of a bird but rather to the movement of the animal, its dynamism in space. The definition adopted by the court in 1916 therefore had to be adapted to the evolution of contemporary art, which certainly allowed one to qualify *Bird in space* as a sculpture.

The main argument of Brâncuși's lawyer (BL) can be schematized as follows:

(P1) All objects that imitate natural objects or represent abstract ideas are sculptures.
(P2) *Bird in space* represents an abstract idea.

(C) *Bird in space* is a sculpture.

How should we evaluate these arguments? As we shall see in a moment, both the premises of CL and the premises of BL provide reasons to support their respective conclusions. So, both arguments satisfy condition (2) of the definition of a good argument. But what about condition (1), which states that the premises of a good argument are true? Which of the two arguments meets this condition? We will give an indirect answer shortly. Before moving on to this problem, let us focus our attention on condition (2).

If we look at the argument provided by CL, a relevant feature of P1 and P2 is that if these premises are true, then the conclusion must also be true. To see this, thinking in terms of sets may be useful. If the set of sculptures is a subset of things that imitate natural objects, then something that is outside the second set must also be outside the first one. If we assume P1, and assume that *Bird in space* does not imitate a natural object (P2), then we must conclude that *Bird in space* is not a sculpture (C).

Premises P1 and P2 of the argument presented by BL have the same feature. In set terms, if the set of what imitates natural objects or represents abstract ideas is a subset of the sculptures, and *Bird in space* is in the first set, then it must also be in the second. These features will help us understand the kind of reasoning that goes under the name of deductive reasoning.

2.3.1 Deductive reasoning

The reasons that CL and BL provide to support their respective conclusions are *conclusive reasons*. In other words, *if* the premises expressing these reasons are true, then the conclusion must also be true. Arguments of this sort are said *deductively valid*. On the contrary, an argument whose conclusion may be false, even though its premises are true, is said invalid or not deductively valid. For the sake of brevity, in this volume we will use the expressions "deductive reasoning" and "deductive argument" to refer, respectively, to deductively valid reasoning and arguments. Here it is a definition of a deductively valid argument:

> An argument is deductively valid if and only if: if all its premises are true, then the conclusion cannot be false.

Two alternative definitions, equivalent to the previous one, are:

(1) An argument is deductively valid if and only if: the conclusion necessarily follows from the premises.
(2) An argument is deductively valid if and only if: there is no logically possible scenario in which the premises are true and the conclusion is false.

These definitions give us an intuitive method to check whether an argument is deductively valid or not. Let us consider CL once again. Is there a possible scenario where P1 and P2 are true, but C is false? No, there is not, and the reasoning in terms of sets proposed above helps us understand why. Instead, if we can construct (think, imagine) a logically possible scenario in which the premises of an argument are true but the conclusion is false, then the argument is not deductively valid.

Two typical schemes of valid deductive arguments are as follows:[3]

Scheme 1	Scheme 2
All P are Q	All P are Q
c is not Q	c is P
c is not P	c is Q

These schemes include so-called "schematic letters". "P" and "Q" stand for predicates, "c" for a name – the name of an individual. In logic, the term "individual" denotes what in most cases we would simply call "an object" or "an entity". Schematic letters are used because the validity of the two schemes outlined above does not depend on which particular predicates and names appear in the premises and in the conclusion.

The validity of the two arguments depends on the logical, formal relations between the sentences serving as premises, on the one hand, and the sentence serving as conclusion, on the other. An example is in order here.

Suppose I say, "All dolphins are mammals". If something (e.g., a given object c) does not have the property of being a mammal, it cannot have the property of being a dolphin. That is, it cannot be the case that dolphins are not mammals (scheme 1). Furthermore, if something has the property of being a dolphin, it will necessarily also have the property of being a mammal (scheme 2). Note that the possibility we are talking about does not depend on the laws of nature, as when I say that it is not possible to reach the moon in one leap. Instead, it is a logical possibility, which depends on the concepts we use in speech and reasoning. That is, it is conceptually impossible for an object to be a dolphin but not a mammal, if we assume that all

---

3. These schemes will be examined in more detail in the fifth chapter of this volume.

dolphins are mammals. This explains why valid deductive arguments provide conclusive, indubitable, incontrovertible reasons to support the conclusion. It is said in this sense that, in the case of valid deductive arguments, the premises *imply* the conclusion, and that it would be inconsistent, given those premises, to deny that conclusion. In general terms, if C necessarily follows from P1, ..., Pn, then we assert a contradiction if we jointly assert P1, ..., Pn and the negation of C.

Let us now return to the Brâncuși case. Which of the arguments provided, respectively, by the customs lawyer (CL) and Brâncuși's lawyer (BL) satisfies condition (1) in the definition of a good argument? That is, which of the two arguments includes true premises only? Neither? Both? CL? BL? We will answer this question indirectly, i.e. without committing ourselves to either CL or BL, but in doing so we will begin to train our reasoning skills.

Both CL and BL are deductively valid: for both, if all their premises are true, then the conclusion must also be true. Let us assume that both arguments satisfy condition (1). This means that the conclusions of CL and BL must be true. But the conclusion of CL is the negation of the conclusion of BL, and a sentence and its negation cannot both be true. So, it is not possible for both CL and BL to satisfy condition (1). Thus, they cannot both be good arguments, even though they are deductively valid.

So what? Of course, it might be the case that both premises of CL are true, in the sense that there are logically possible scenarios in which they are so. In those scenarios, the conclusion of CL is also true. The same can be said for BL (remember, however, that no scenario that makes premises of CL true makes premises of BL true, and vice versa). So, there are logically possible scenarios in which CL is a good argument (and BL is not), and scenarios in which BL is a good argument (and CL is not). There are also scenarios in which neither is a good argument – for example, scenarios in which P1 of both arguments is false.

All that being said, which one of the two is a good argument? According to the judges who decided the Brâncuși case, BL was a good argument and CL was not. Indeed, on the basis of the evidence gathered during the trial, the judges came to the conclusion that all the premises of BL were true, whereas one of premises of CL was false. This allows us to anticipate a point that we will discuss in more detail in Chapter 4. If we look at the definition of a good argument, there is an important difference between the first condition (truth of the premises) and the second condition (the premises provide reasons for the conclusion). In order to see whether condition (1) is fulfilled we have to see how things actually are. This can be a rather complicated matter, and perhaps, in some cases, a matter that we cannot ascertain at all. Even in those cases in which the truth-value of the premises can be found, the individual cognitive agent may not be in a position to ascertain how things actually are. That is, they may not have sufficient elements to believe that all the premises of an argument they are presented with are true, or to believe that at least some of them are false. Between us and the way things are, in short, there is a filter: that of our reasonable and justified beliefs about the reality around us.

The filter we are talking about is epistemic in nature.[4] It depends on what we believe, at least after adequate exploration of the world, and on how much we know. In short, truth is not transparent to us. We need to think and ascertain the truth, and in doing so we may get it wrong, and end up believing something false, or not believing something true. This explains why, in certain situations, we may not be able to tell, at least without proper investigation, whether the premises of an argument are true or false.

To determine whether condition (2) is satisfied, on the contrary, we do not need to look at how the world is. Giving reasons does not depend on how things are; it depends on the relations between the sentences that serve as premises and the sentence that serves as a conclusion. That is why we could say that CL and BL are deductively valid. What is required here is to understand that if the premises are true, then the conclusion must be true. To see this, we do not need to know or believe that the premises are true.

However, the truth of the premises is important in our reasoning, to such an extent that we are often busy trying to ascertain or disprove it. The following definition gives us a technical notion, that of a sound argument:

Definition of a sound argument
An argument is sound if and only if:
(1) all its premises are true;
(2) it is deductively valid.

If (2) is satisfied, but (1) is not, then we have an argument that is deductively valid but not sound. From the definition of a deductively valid argument, and from the definition just given, we have that the conclusion of a sound argument is true.

In the rest of this volume, we will use the expression "sound argument" in this technical sense, thus departing from common usage. When we speak of "sound argument", therefore, we shall be speaking of deductively valid arguments having true premises. What we are now going to look at will make it clear that, in this technical sense, "sound argument" is not synonymous with "good argument".

2.3.2 Good arguments and deductively valid arguments

It is now worth focusing on the relationship between deductively valid arguments and good arguments. Can there be deductively valid arguments that are bad arguments? The answer is affirmative. We saw earlier that at least one of AD and BL is not a good argument, although both are deductively valid. More generally, it is easy

---

4. The term "espistemic" comes from the ancient Greek word *episteme* (knowledge) and indicates all that has to do with knowledge. Therefore, Epistemology is that branch of philosophy whose aim is to define and clarify our concept of knowledge.

to find cases of arguments that are deductively valid and yet have at least one false premise – and are therefore not good. Consider the following example:

(P1) All men live forever.
(P2) Socrates is a man.

(C)   Socrates will live forever.

There is no logically possible scenario in which P1 and P2 are both true and C is not. Thus, the argument above is deductively valid. But P1 is false, and so the argument is not a good argument. This can happen because validity is a characteristic that concerns the form of deductive arguments, i.e. the relation between premises and conclusion, regardless of whether they are true (or false). In other words, an argument is valid if it preserves the truth. If the premises are true, then the conclusion cannot but be true. But validity in itself does not tell us whether the premises are true or false.

What if an argument is deductively valid and has only true premises? Is it necessarily a good argument? To put it differently, can there be sound arguments that are not good arguments?

This time the answer is more subtle. In general, sound arguments are good arguments. There is, however, one type of deductively valid argument that cannot be good, even when it has true premises. These are *circular arguments*, which we will consider in Chapter 4, where we will see why they cannot be good arguments. Circular arguments, implicitly or explicitly, contain the conclusion among their premises. So, although they are deductively valid (if their premises are true, then the conclusion is also true), and although their premises may be true (and so they are sound), circular arguments do not provide reasons for their conclusion because they presuppose what they are supposed to prove. Apart from circular arguments, however, sound arguments are good arguments.

We have seen an example of a deductively valid argument with at least one false premise. In the light of the definition of a good argument, this is a bad argument. There are also cases in which an argument is bad even if it has true premises. This is what happens to arguments in which the premises do not provide reasons for the conclusion. Here it is an example.

In the course of the Brâncuși trial, the court called experts in the field of art to testify, including the sculptor Robert Aitken. Aitken acknowledged that during the last two decades the way of conceiving art had changed. Evidence of that was given by the fact that many sculptures, contrary to the 19th century tradition, did not imitate natural objects but rather represented abstract ideas. However, Aitken continued, *Bird in space* did not please him and was therefore not a sculpture. Let us try to schematize Aitken's argument:

(P1) All sculptures imitate a natural object or represent abstract ideas.
(P2) I don't like *Bird in space*.

(C)   *Bird in space* is not a sculpture.

Now, the premises of this argument are true, but the conclusion does not follow from the premises, since the premises do not give any support to the conclusion. As Aitken himself implied, there is no relationship between the fact that I like something and the fact that that something is a sculpture.

2.3.3 Kinds of non-deductive reasoning

At this point, we can answer a question that should come quite naturally: can we have good arguments that are not deductively valid? The answer is affirmative. There are many kinds of reasoning that we use in various important contexts – including, for example, legal disputes and scientific research – in which we use arguments that start out from true premises, provide reasons to support the conclusion, and yet are not deductively valid.

During the Brâncuși trial, the sculptor Jacob Epstein was called to testify, as an expert in the field, as to whether *Bird in space* was a sculpture. After reviewing his credentials as an artist, the judges asked Epstein whether he knew Brâncuși and considered him to be a sculptor in his own right. Epstein said that he had known the Romanian artist and his works for many years, works that were appreciated everywhere for their originality and aesthetic quality. The judges then asked Epstein if he considered *Bird in space* a work of art on a par with the previous sculptures made by Brâncuși. Epstein answered in the affirmative: that work had clear artistic qualities; moreover, if Brâncuși claimed that *Bird in space* was a sculpture, there was no reason to believe otherwise, given the artist's reputation. We can reconstruct the last passage of Epstein's argument as follows:

(P1) Brâncuși is a famous sculptor because he created many highly regarded sculptures in the past.
(P2) Brâncuși claims that *Bird in space* is a sculpture.

(C) *Bird in space* is a sculpture.

Is this argument good or bad? Its premises are certainly true, but they do not provide conclusive reasons in favor of the conclusion being true. Verifying the latter point – namely, that the argument in question is deductively invalid – is easy. Consider the following logically possible scenario: Brâncuși, a well-known sculptor, wants to create a designer juicer to sell on the market; dissatisfied with the result, he then claims that it is a sculpture, called *Bird in Space*, exploiting his fame to deceive a possible buyer. In this scenario, P1 and P2 are true, but C is not.

Based on this, should we conclude that Epstein's argument is not a good one? It depends. The argument seems to assume that normally (or by default) a well-known artist does not deceive about the status of what he creates. Therefore, we can assume he is telling the truth about it, unless we have specific reasons to think otherwise.

The principle is quite reasonable. A famous artist would not risk his or her reputation by passing off, say, a poorly made juicer as a sculpture. What if the rest of the art community refused to bite and did not discern anything artistic about the object?

If we consider the principle in question compelling enough, then Epstein's argument is a good one. To say that we consider the principle "compelling enough" is to say that we do not need a principle that makes the falsity of C even more improbable or implausible in the presence of the truth of P1 and P2. If the default principle presupposed by Epstein's argument is compelling enough for us, then P1 and P2 provide reasons to support C.

There are several types of non-deductive reasoning that can produce good arguments. The most common are induction (inductive reasoning), abduction (abductive reasoning), and analogy (reasoning by analogy). They perform different functions in human argumentation and reasoning, as we will see in the next chapters. What they have in common is the following feature: all good arguments that are non-deductive in form provide non-conclusive reasons for believing that the conclusion is true. The truth of their premises somehow makes it more plausible, or more likely, that the conclusion is true rather than false, but nothing more, because in this kind of reasoning the conclusion does not necessarily follow from the premises. In short, inductive reasoning, abductive reasoning, and reasoning by analogy amount to argumentation schemes that are non-deductively valid, but this does not imply that they are not good arguments.

There are criteria other than validity for evaluating how well a non-deductive argument works (how good it is). In general, a non-deductive argument is strong if its premises provide a high degree of confidence in its conclusion. Otherwise, an argument is weak. In other words, a strong argument will, under the assumption that all its premises are true, make it very likely, or plausible, or reliable that the conclusion is true. A weak argument will fail, even under that assumption, to increase the probability of the conclusion being true. Even in the face of a strong argument, however, we cannot rule out the possibility that the conclusion is false.

An example of this is provided by the story of the inductivist turkey proposed by the British logician and philosopher Bertrand Russell (1872-1970). This famous story clarifies why inductive reasoning, no matter how well supported it is by its premises, can always lead us to a false conclusion. A turkey living on a farm decided one day to elaborate a scientific view of the world based only on observable facts. The turkey observed that every day, at 9 am, food was brought to him, and this happened whether it rained or it was nice weather, in summer and in winter, on weekdays and on holidays. On the basis of the observations accumulated over time, the turkey came to the conclusion that at 9 o'clock every morning he will be given food. A conclusion that proved to be indisputably false on Christmas Eve, when, instead of being fed, the turkey was slaughtered. The turkey had used a strong non-deductive argument concluding that he would be given food every morning at 9 am, but, nevertheless, his conclusion was false. It goes without saying, however, that non-deductive arguments are very important in everyday life as well as in the development of knowledge. Yet, they should be handled with caution, without expecting from them the certainties they cannot give us.

One might wonder why we do not always make use of deductively valid arguments, given that they warrant that the conclusion is true, and not simply that it is probably true. As we shall see, it is not always possible to support a given conclusion by means of deductive reasoning. As a consequence, we must often make do with non-deductive reasoning. Moreover, deductive reasoning, as opposed to non-deductive reasoning, generally does not expand our knowledge, in the sense that its conclusions do not increase the information contained in the premises, except in a trivial or irrelevant way. In other words, deductively valid inferences are not ampliative. This is because, in a deductively valid inference, the information provided by the conclusion is, at least implicitly, already included in the premises. Let us try to illustrate this with an example. Consider the following inference:

(P1) All men are mortal.
(P2) Socrates is a man.
___
(C)   Socrates is mortal.

The inference from (P1) and (P2), taken together, to (C) is deductively valid. The information provided by the conclusion, however, adds nothing to what is already said by the premises.

The fact that deductively valid inferences are not ampliative makes us understand why we routinely need other kinds of reasoning when we have to explain surprising phenomena. For in this case, we want to get an information that is not already provided by the evidence available to us, which serves as a premise for our explanations. For this reason, when we need to provide explanations of phenomena for which we do not know the reasons or causes, we resort to forms of reasoning that are not deductively valid.

For example, inductive inferences, although fallible, expand our knowledge:

(P1) All ravens that have been seen so far are black.
___
(C)   The next raven we see will be black.

This is an example of induction by enumeration. Two considerations immediately jump out from this example. An inference of this type is ampliative: the information provided by the conclusion is not contained in the premises. For example, suppose that $r_{100,000}$ is the next raven we see, and that r_1-$r_{99,999}$ are the ones we have observed so far. The information that $r_{100,000}$ is black is, of course, not contained in the information that r_1-$r_{99,999}$ are black.

Second, this inference is fallible. We might discover in the next observation that $r_{100,000}$ is white, despite the fact that the inductive argument above is good.

Another feature of inductive inferences is that they are non-monotonic, i.e., adding new premises to an argument can lead us to reject the conclusion of the argument. For example, consider the following inference:

(P1) r_1–$r_{9,999,999}$ are black.

(C) All ravens are black.

Suppose now, for the sake of the argument, that raven $r_{10,000,000}$ is not black. By adding to (P1) the premise (P2) "$r_{10,000,000}$ is not black" we are no longer in the position to conclude that "All ravens are black", despite the fact that, prior to the addition, the inductive argument was good. Conversely, deductive reasoning is not ampliative, and therefore cannot replace induction or abduction in cases where we want our conclusions to give us information that is not already present or implicit in the premises. The point is precisely this: on the one hand, the inability to expand our information is the price we must pay for being sure that the conclusion is true (if the premises are also true), and on the other hand, fallibility is the price we must pay for expanding our knowledge.

Since we want our explanations of facts to be ampliative, it follows that fallibility is the rule in each of our attempts to explain facts. Is it worth rejecting induction and abduction because they may lead us to false conclusions, or because they may force us to abandon the conclusions we had previously drawn? If that is the case, we should also reject our activity of explaining facts and, with it, a huge portion of scientific inquiry. In other words, we cannot give up using inductions and abductions. They are extremely useful from the epistemic point of view.

2.3.4 Strength and weakness of non-deductive reasoning

We have seen that non-deductive arguments, i.e. those arguments that do not express deductive reasoning, can be qualified as "strong" or "weak": they are strong when providing a high degree of confidence in their conclusion, week when this is not the case. This informal characterization is in need of closer scrutiny.

First, the informal characterization in question speaks of "degrees of confidence". The greater the confidence provided, the stronger the argument. So, on this account the strength or weakness of an argument is a matter of degrees on a continuous scale. This makes this distinction very different from that between deductively valid and invalid arguments. According to the latter distinction, an argument is either valid or it is not: it cannot be more or less valid. On the other hand, cases of non-deductive reasoning can be stronger or weaker. There is no sharp cut-off point here: some arguments are extremely strong, some are extremely weak, and some are in a borderline situation.

Second, the characterization in question makes reference to a "high degree of confidence". If we want to somehow measure the degrees of confidence, which values make a degree "high", and which do not? The answer here is rather complex. To cut a long story short, we can say that it is not possible to establish *a priori* and in the abstract which degrees of confidence are high enough to generate arguments that are sufficiently strong. The threshold we choose depends on a number of pragmatic and contextual elements, which include the type of activity in which our arguments are involved. For example, a default principle such as the one presupposed by Epstein's argument may be quite stringent in the context of a legal dispute, where we consider relationships between facts that rarely have precise measurements (there are no statistics about how many artists lie about their creations). In a scientific context, we may need more stringent criteria, at least quantitatively – that is, we may need arguments that assume precise and statistically very incisive quantitative relationships between two phenomena or properties.

That said, we should not confuse the strength of an argument with its effectiveness. As just noted, an argument is strong when its premises warrant the conclusion to a high degree. On the other hand, an argument is *effective* when it is able to convince other people to believe in the conclusion. On many occasions it happens that bad arguments prove rhetorically effective, and good arguments prove rhetorically ineffective. Put differently, providing a good argument is not always enough to convince other people to believe in a certain conclusion. This is because people's opinions are often affected by prejudices, false beliefs, errors in reasoning, and emotional states. All this leads human beings to accept a claim even though it turns out to be unacceptable or implausible when subjected to the scrutiny of a good argument.

2.4 Conclusion

Let us recapitulate. We have seen that a good argument is endowed with two characteristics: the premises are true, and they provide reasons to support the truth of the conclusion. An argument is bad in the case that it is not endowed with one or both of these characteristics.

With respect to their argumentative structure – that is, the relationship between premises and conclusion – arguments can be distinguished into two families: those that express deductive reasoning and those that express non-deductive reasoning. Deductive arguments are used to provide conclusive reasons because they warrant that if the premises are true, then the conclusion cannot be false. If this condition is not met, an argument is said to be deductively invalid. Nevertheless, nothing precludes the premises of a valid argument from being false, as well as the premises of an invalid argument from being true. The validity of an argument is a purely formal characteristic of it, which is completely independent of the truth of the premises. The non-deductive reasoning instead does not give conclusive reasons for the conclusion, but reasons to consider it probable, plausible, likely, reliable. For them it holds that if the premises are true, we have reasons to believe that the conclusion is not false, although this possibility can never be completely excluded. In this sense we say that a non-deductive argument is (more or less) strong or (more or less) weak depending on the degree with which it warrants the conclusion. Thus, a non-deductive argument will be good if its premises are true and if it provides strong reasons to support the conclusion.

Finally, the strength of an argument should not be confused with its effectiveness, that is, its ability to convince other people to believe in the conclusion. It frequently occurs that bad arguments prove to be very effective in everyday life and in public debate because of the prejudices, false beliefs, or cognitive biases of their recipients.

These observations allow us to highlight once again the normative function of the study of argumentation. This study makes it possible to identify criteria for distinguishing good from bad argumentation. This is very useful when we are called upon to evaluate other people's arguments and to detect errors in reasoning, particularly in contexts where bad arguments are more effective than good ones.

3 | Rational discussion and the pyramid of disagreement

In the previous chapter we learned to identify an argument. In this chapter, we will deal with the ways in which disagreement can be expressed with respect to an issue. There are rational and non-rational ways of expressing disagreement. A disagreement based on reasons, and therefore guided by a rational attitude, allows a better understanding and refutation of others' view. Knowing the rules governing a disagreement based on reasons is therefore very important. We will discuss some properties that an exchange of reasons must have in order for a discussion to be rational (§ 3.2), examining the so-called "rules of rational discussion" proposed by the scholars of argumentation theory Frans H. van Eemeren and Rob Grootendorst. We will then move on to consider different ways of expressing a disagreement within a discussion, beginning from the least rational (harmful or useless for the understanding of the point at issue) to the most rational (§ 3.3). The ability to evaluate an argument – that is, to understand if it is good or not – is fundamental. Without it, we would accept (or refuse to accept) an argument in a completely arbitrary way. Furthermore, without such a skill we could not rationally reply to an argument.

3.1 Introduction

Usually, in a rational discussion, the evaluation of an argument is aimed at the elaboration of a reply. In other words: we try to judge an argument because we want to see if some of its features make it bad, and therefore susceptible to criticism. Obviously, the situations in which a subject elaborates a reply to an argument are those in which she disagrees with the proposed argument. Precisely for this reason, in the second part of this chapter we will focus on the possible ways in which disagreement is expressed. There are two major types of ways by which disagreement can be expressed: rational – which help us to better understand whether the premises of the argument really give reasons to support the conclusion, and whether the premises are true – and irrational, which do not help us to understand the other's claim or the reasons for it. In talking about irrational ways of replying to an argument, most texts introduce fallacies – i.e. erroneous schemes of reasoning – and systematically discuss different types of them. In this volume, we will proceed differently. We will look at some fallacies in § 3.3, and others in the next chapters of the volume. We have chosen not to write a chapter devoted to fallacies, detached from the rest of the volume, but to discuss specific fallacies relating to the types of reasoning we will address throughout the chapters of this book. Furthermore, we will not present the taxonomy of fallacies that was established by Aristotle and, later, by medieval and modern logicians. Instead, in this chapter, we will discuss the so-called pyramid of disagreement proposed by Paul Graham, who divides ways of reacting to disagreement into six types: four clearly irrational, and two more rational, of which only one, however, is a fully rational reply to an argument.

3.2 Rational discussion

A discussion is a verbal or written interaction between two or more subjects. More precisely, when a discussion arises, someone puts forward a claim and someone else reacts by expressing agreement or disagreement with her. If the disagreement is supported by an argument, we have some form of reply to the original argument. We will focus on the latter aspect especially in § 3.4, but it also plays a role in what we will say in the next few paragraphs.

There are various reasons why we advance an argument or reply to it. For example, we can do this because we want to persuade an audience or interlocutor. In this case, it is not the goodness of the argument that matters, but its effectiveness – its success in persuading our interlocutors. Or, we can propose an argument or reply to it because we want to obtain an adequate understanding of the claim that is a source of disagreement, and we want to inquire into the rational link between the premises and the conclusion that expresses the claim. In this case, what matters is not persuasion, but the goodness of the argument, as we will see shortly. Obviously, the two aims are compatible: we can discuss an argument aiming to adequately understand the point at issue, and at the same time persuade our audience, including those who initially oppose our position. In that case, we aim at something that is not mere persuasion but rational agreement. However, it is also possible that persuasion and goodness of an argument do not go together. As we saw in Chapter 2, an argument can be persuasive even if it is not good, and conversely, a good argument can be very unconvincing.

In this chapter, and more generally in this volume, we focus on what distinguishes good (counter-)arguments, while we do not discuss the persuasive aspects of communication. This is for two reasons. First, only an adequate understanding of what the discussion is about can help us take a well-founded position on what is being discussed. This is particularly important in public debates, where the discussion is about important individual or collective choices, but it is also relevant in other cases. In general, each of us has a pragmatic interest in understanding certain topics, because a good understanding is a prerequisite for making rational decisions. Second, we are interested in knowing truth, and in distinguishing it from falsehood. This interest can be purely theoretical, or once again it can be for pragmatic reasons: knowing whether something is true or not can be relevant to our decisions.

In short, we are interested in those discussions that can give us a better understanding of the point at issue and of the reasons in favor of its truth. This is important given that most of our cognitive and intellectual activities are (or presuppose) truth-seeking activities (think about scientific research, for example).

For simplicity we will assume that subjects, whether rational or not, are sincere in the course of a discussion – that is: a subject only asserts something if she believes what she is asserting.

3.2.1 The rules of rational discussion

To be rational, a discussion must respect a certain number of features – in general, it must satisfy some requirements to ensure that it improves our understanding of the point at issue. These features are commonly referred to as "rules of rational discussion". Different argumentation theorists propose different sets of rules. Here we will focus our attention on some of the rules proposed by van Eemeren and Grootendorst,[1] because they are particularly relevant to our purposes.

1. *The participants in a discussion must not prevent others from putting forward a claim or raising objections.* According to this rule, therefore, all people who participate in a discussion have the right both to put forward a standpoint and to object to the standpoint of the other participants.
2. *If someone attacks someone else's claim then their attack must be directed against that claim.* In a rational discussion, one should attack the views proposed by another participant, not the person proposing them. Furthermore, it is necessary to specifically attack the position of the other participant and not a different position that the other participant has never advanced and that is falsely attributed to her.
3. *Whoever puts forward a claim should defend it if others request this.* According to this rule, if the other participants express doubts about our position, we are required to provide some reasons in support of it. There are, however, restrictions on this rule. For example, I need not answer an objection if I have already successfully answered that objection. A second case in which this rule does not apply is when the objector does not respect the rules of rational discussion, for example violating the first two rules.

---

1. Frans H. van Eemeren, Rob Grootendorst, *A Systematic Theory of Argumentation. The Pragma-dialectical Approach*, Cambridge University Press, Cambridge 2004. See also Frans H. van Eemeren, Rob Grootendorst, *Argumentation, Communication, and Fallacies*, Routledge, London and New York 2016 (first ed. 1992). Van Eemeren and Grootendorst propose 10 rules in their work. We will discuss only 7, following an order different from that followed by these scholars.

4. *You can defend your own claim only by putting forward an argument in favor of it.* As we said in § 2.1, in rational discussions, the only way to defend one's opinion and to convince others is to put forward an argument in favor of that opinion. No other forms of defense are permitted.
5. *The premise of an argument should not be presented as a position already accepted by all, nor should one deny a premise that has already been accepted during the discussion.* This rule prescribes two things: a) not only does everyone have the right to object to a claim, but everyone also has the right to call into question one or more premises of the argument that other participants put forward in defense of their claim; b) if, however, someone has already defended her claim with an argument, and that defense has been accepted by the other participants, then the other participants can no longer question that claim when it is used as a premise for another argument.
6. *A claim can be considered successfully defended only if valid or sufficiently strong arguments have been used in its favor.* Therefore, if an opinion has been defended by invalid or weak arguments, the other participants always have the right not to accept it.
7. *If a claim has been successfully defended, then others must withdraw their doubts regarding that claim. If the defense of a claim has not been successful, then whoever has advanced it must withdraw it.* Therefore: a) if a claim has been advanced by means of sufficiently good arguments, and if the attacks aimed at the premises of those arguments have not been successful, then the other participants must accept that claim; b) conversely, if the objections to a claim have been successful, then whoever advanced that claim should drop it.

These rules are *normative* and not *descriptive*: *they tell us how to conduct a discussion or exchange of views, not how we actually take part in the activity of discussing a claim.* Our daily experience certainly gives us frequent examples of these rules being violated. In particular, people tend to be very attached to their opinions and to identify with them. This has two consequences: on the one hand, people often find it difficult to drop their position even in the face of a considerable number of reasons to the contrary – therefore they find it difficult to follow rule 7. Even when they do not successfully defend their position, they do not withdraw it because they consider it part of their identity.[2] Second, people often understand an attack on their opinion as an attack on their person, even when it is not (that is, even when rule 2 is not violated). However, if the criticism is directed at the defended claim, and therefore respects rule 2, it is incorrect to react as if it were a personal offense. This favors subsequent violations of some of the seven rules (perhaps in particular of rule 2) by those who fail to see the objection as a reply to their argument. To avoid this, it is import-

---

2. This phenomenon has recently been highlighted by Michael P. Lynch, *Know-It-All Society: Truth and Arrogance in Political Culture,* Liveright Pub Corp, New York and London 2019.

ant to be able to overcome the seemingly natural or at least frequent tendency to identify with one's own opinions, be they political, religious, football or other. The same error is often committed by those who attack the opinions of others: they often do not distinguish the position that the other participants are defending from the persons who defend them and tend to identify the two. This often results in the violation of the first two rules: instead of showing that the interlocutor's opinion is wrong by means of rational arguments, we end up questioning his person. This causes us to stray beyond the perimeter of rational discussion. As we will see in the next section, this exit leads to a stalemate, in which everyone stands on their own position and no progress is made towards resolving the dispute. Furthermore, this fuels the conflict and thus banishes the search for shared views.

3.3 Ways of reacting to disagreement

In the context of a rational discussion, one replies to an argument (as well as to a claim) when one disagrees with it. There are many ways of expressing disagreement, and as will be clear from the examples we will discuss here, not all of them are rational – that is, not all of them abide by the rules we have listed above. This topic and its implications are discussed in a famous article by the computer scientist Paul Graham. Shortly after the mid-noughties, Graham was the first to observe how blog and forum users actually conduct an online discussion, and especially how they react to a claim or argument proposed by one of the participants in the discussion. In *How to Disagree,*[3] Paul Graham distinguishes seven forms or levels of disagreement within a conversation, which can be ordered in a pyramid that goes from the least rational ways of reacting, in the lower levels, to the most rational ones, in the highest. For the sake of simplicity, we will consider only six levels here. According to our characterization of the rationality of a reply, this means that the reactions listed in the lower levels do not make any contribution to the understanding of the claim initially proposed or of the reasons in its favor. However, this contribution increases as one gets to the next levels, until one reaches the top level, i.e. the best possible contribution to the discussion, and therefore the most rational way of reacting. Let us analyze the possible ways of expressing disagreement using an example. Suppose a social contact of ours writes on her Facebook wall:

> Banning same-sex marriage is unacceptable because it constitutes a form of discrimination.

Before considering the possible reactions to this statement, it should be noted that not only does our contact's post support a claim ("Prohibiting same-sex marriage is

---

3. http://www.paulgraham.com/disagree.html.

unacceptable"), but also advances a reason for this ("Prohibiting same-sex marriage constitutes a form of discrimination"). For the argument to work, it is necessary to add a premise that seems implicit in the context of the post: "Any form of discrimination is unacceptable". Note that our social contact had a good reason for omitting this last premise: in any constitutional state (as we believe ours is), it is a common belief that any discrimination is unacceptable, and thus this premise can be presupposed.

Let us now turn to the possible ways of reacting to the argument – more precisely, to those listed in the pyramid of disagreement. Suppose a comment on the post reads: "You're a pervert!", And another: "You homosexuals should be locked up". This is the most irrational mode of reaction, represented by level 0 (the lowest) in Figure 3.1, and consists of insulting one's interlocutors or threatening them in some cases. This modality clearly contravenes some of the seven rules that we have seen in § 3.2. It is obvious that it violates rule 2, but on closer inspection it also violates rule 1, because the insult has the effect of discouraging the other interlocutors from expressing their opinion, if not intimidating them. Furthermore, this highly irrational mode of reaction prevents the conditions under which rules 3 to 7 can be satisfied, since it does not permit a rational discussion of the claim the social contact has proposed. More generally, the insult is the most irrational way of reacting because it contributes least to the understanding of the claim initially advanced and of the reasons that support it.

Figure 3.1 The pyramid of disagreement by Paul Graham

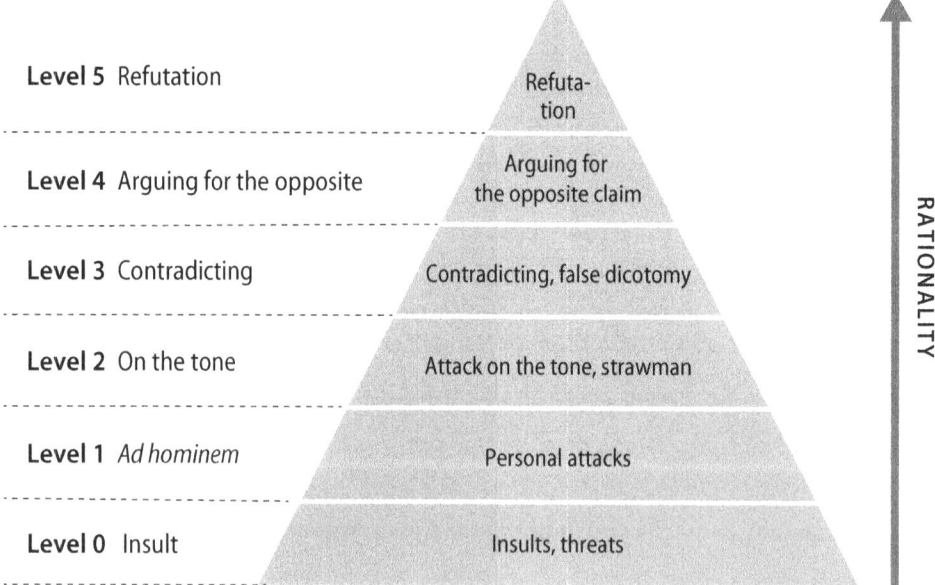

Source: Elaboration from P. Graham, *How to Disagree*, 2008, http://www.paulgraham.com/disagree.html.

In the traditional classification of the errors of reasoning (fallacies), when a standpoint is answered with threats, it is said that an *argumentum ad baculum* (literally "argument of the stick") is used: to convince others, instead of proposing an argument, one threatens them.

Now suppose that another comment reads: "You are a member of the gay lobby". This is an example of reaction of level 1: the personal attack, also known as *argumentum ad hominem* (lit. "argument at the person"). This kind of reaction targets the interlocutor by insinuating that they are supporting their claim not for the sake of truth but because they only wish to defend personal or group interests. Implicitly, this suggests that the proponent of the claim does not have the credibility or the authority to do so – without saying anything about the claim. It is clear why this reaction is also very low in the pyramid: again, by reacting in this way, the contact's claim is not even considered, and therefore no contribution is made to the understanding of this claim or of the reasons that support it. Note that many forms of personal attacks (such as the one in our example) are offensive to those who put forward the initial claim, just like the insults, albeit in a more subtle way. They actually constitute a form of verbal violence, whose effect is to dissuade others from participating in the discussion.

It should be noted that objections grounded on the inconsistency between the argument and the behavior of those who support it also fall into this level. Suppose a smoker claims that smoking is bad and that one of his interlocutors says: "How can you claim that smoking is bad when you are a smoker?". Now, this reply is not offensive, and may have a minimal rational basis: if the person proposing the argument behaves as if the conclusion were false, this leads us to suspect that the argument is not good or that the proposer is not sincere and, for some reason, is propagating falsehoods. However, what really matters in a rational discussion is the consideration of the proposed argument: the robustness (and truth) of its premises, the truth of the conclusion, and the presence of adequate rational links between them. Moreover, in principle it is possible that whoever proposes an argument is behaving as if the argument were not good even though this is not the case. The attitude of the person proposing an argument towards the conclusion reached is not always a reliable indicator of the goodness of the argument. Note that a similar consideration is also valid in the case of a biased interlocutor: an argument can be good even if it is proposed by a person who has an interest in convincing us that the conclusion is true, and must therefore be analyzed and discussed independently of this aspect.

In any case, personal attacks violate rule 2 of rational discussion, which requires that one should attack the claim under discussion and not the proponent. This is also true in the example of the inconsistent smoker: the reply above does not consider the argument that the smoker has proposed (whatever it is), but only this person's behavioral inconsistency. This calls into question their credibility without at the same time improving our understanding of the reasons put forward against smoking.

The next way of reacting, indicated by level 2, is "to reply to the tone". Suppose that another Facebook user comments on our contact's post by writing: "I can't believe such a thing is said so lightly". This is an example of the kind of reaction in question. By replying in this way, it is questioned whether the claim was posed properly or with all due seriousness. In this case too, rule 2 of rational discussion is violated, as the argument is not attacked but only its tone is. More generally, just like the previous reactions, this reaction does not take us a single step towards a better understanding of the point at issue. Before considering the next level of the pyramid, let us discuss a way of reacting that Graham does not consider. This is the *straw man fallacy*, which consists in attributing to our interlocutor conclusions that they have not defended or in attributing to them an argument that they have never proposed. Suppose that another comment to our post reads: "By saying this you want to destroy the very institution of the family, but this is wrong". The person who put forward the initial argument did not actually come to this conclusion. In order to argue that the extension of the institution of marriage to persons of the same sex implies the end of the family institution, one should advance another argument. Obviously, if the discussion is directed at something which has not been proposed, we are prevented from enhancing our understanding of the point at issue – indeed, we risk losing sight of it. It is therefore easy to understand why the straw man is not a rational way of disagreeing. Why do we often react this way when we disagree with others? Usually because the claim that we wrongly attribute to our interlocutor is weaker and more attackable than the one that they actually put forward. However, rule 2 prescribes that the target of our reply should be the claim that our interlocutor has actually advanced and not something that – often remotely – resembles it.

Level 3 is the way of reacting that Graham calls "contradiction", and which we prefer to call the "act of contradicting", for a reason that we will clarify shortly. This reaction to disagreement consists in affirming something that implies the negation of the claim supported by the interlocutor. The extreme case of this kind of reaction simply consists in denying the claim advanced by the interlocutor. Suppose that one of the comments to our example post is: "It is not unacceptable to prohibit same-sex marriage", or "It is unacceptable to allow same-sex marriage", and suppose that the comment adds nothing else – that is, it does not present reasons in support of the denial of the claim in the post. Consequently, in this case too, the disagreement is not substantial since it does not concern the *reasons* used to support a claim but merely denies it. Therefore, the simple contradiction, alone, violates rule 3 of rational discussion (which, we recall, prescribes that a standpoint should not be simply asserted but should also be defended by arguments). More generally, merely denying the initial claim, or something that it implies, does not help us to reach a better understanding of what our interlocutor is saying and of the underlying reasons.

Since it presents no reasons against the initial claim, simple contradiction is an attempt to oppose a claim by presenting the contrary opinion as self-evident, even when, in fact, it is not. An analogous attempt, similar to the act of contradicting, consists in placing the burden of proof on the interlocutor. Suppose, for example, that I claim that God exists and, when someone asks me to prove it, I say: "I'm not the one that must prove that God exists, you are the one that must prove that God does not exist". In this way, I avoid the burden of proving what I claim by placing on the other the burden of proving the opposite claim. Of course, this does not take us one step towards a more structured understanding of the reasons for or against God's existence. The important thing here is to establish on which side the burden of proof lies.

Graham calls level 3 "contradiction", but we prefer to call it the "act of contradicting". There is a precise reason for our choice. A contradiction is a sentence, and more precisely a conjunction whose conjuncts are one the negation of the other.[4] For example, "Any form of discrimination is unacceptable, and some form of discrimination is acceptable" is a contradiction. Contradictions cannot be true: a conjunction is true if and only if both conjuncts are true, and yet if a sentence is true, its negation must be false, if the Aristotelian law of excluded middle is presupposed. By contrast, the act of contradicting is an act: it is something we *do* when we assert the negation of what has been affirmed by our interlocutor, or something that such negation implies. It is not a sentence. Since Graham actually refers to the act of denying a claim when characterizing level 3, we prefer to call this way of reaction "act of contradicting". The relationship between contradiction and the act of contradicting is then clear: a sentence and its negation cannot both be true, but one of the two must be. So either the claim proposed in a discussion is true, or its denial is true. The assertion of a sentence is an act by which one commits oneself to the truth of the sentence (and to the falsity of everything that denies it). So, if I assert something that denies a claim, I am implicitly arguing that that claim is false. Its relationship with contradiction accounts for the dialectical role played by the act of contradicting. It should be borne in mind, however, that the simple act of contradicting is not strictly rational, because the simple act of contradicting does not, in itself, make any substantial contribution to the discussion.

One way of reacting that Graham does not consider is the *false dichotomy,* which is however very common and therefore deserves attention. The false dichotomy consists in proposing a false alternative between the acceptance of one claim and that of another, usually clearly false or in any case having unacceptable consequences. For example, if someone replies to the initial claim of the post: "Either same-sex marriage is forbidden, or the destruction of the institution of marriage is inevitable", then a false dichotomy is being proposed. The fallacy consists in presenting the two alternatives as if they were the only possible ones, when this is not

---

4. In classical logic, which we follow in this book, a contradiction is better defined as "a sentence that implies any other sentence" or as "a sentence that cannot be true". In any case, such sentences are equivalent to the conjunctions we mention in the main text.

the case. In fact, there is no conceptual link between the extension of an institution to a previously excluded category and the end of that institution. Other examples of false dichotomy are: "Anna criticized socialism. She must be a fascist"; "Giulio criticized the reducing of the tax burden. He is certainly a communist".

The ways of reacting to disagreement that we have seen so far are arranged in the Graham scale on different levels, where level 3 is less irrational than all the previous ones, level 2 less irrational than levels 1 and 0, and so on. At the same time, these levels represent inadequate ways of reacting, because they do not make any contribution to understanding the point at issue or the reasons for or against it. In the higher levels of the pyramid we find more and more rational ways of reacting (even if not all of them turn out to be wholly adequate). Let us consider them more closely.

At level 4 of the pyramid we find the reaction called *arguing for the opposite*. Starting from the base of the pyramid, this is the first type of reaction in which the disagreement becomes significantly rational: those who object do not limit themselves to denying what the other says but provide *reasons* for their position. In this case an *argument in support of the opposite claim* is proposed. Let us consider this with the aid of an example. In one of the comments on the wedding post we read: "Same-sex marriage is unacceptable because the family is made up of a man and a woman".

What can we notice here? First of all, the comment contains an argument which works only if we add further premises, namely "There is no acceptable family model other than that consisting of a man and a woman", "Marriage automatically creates a family", as well as "What has unacceptable practical consequences is itself unacceptable". Let us reconstruct the argument "in complete form":

(P1) The family is made up of a man and a woman.
(P2) There is no acceptable family model other than that made up of a man and a woman.
(P3) Marriage automatically creates a family.
(P4) What has unacceptable practical consequences is itself unacceptable.

(C) Same-sex marriage is unacceptable.

The conclusion of this argument contradicts that of our contact's post. Note that the argument thus reconstructed is deductively valid, exactly like the one proposed initially. Accepting its premises is therefore sufficient for rationally accepting the conclusion, and thus for dropping, or simply not accepting, the conclusion of the initial argument. In other words, if the commenter's (reconstructed) argument is sound, then the initial argument cannot be sound.

We are therefore in a situation similar to that of the Brâncuși case we looked at in Chapter 2, but with a crucial difference: the artist and customs lawyers were then asked to argue in favor of their respective premises – that is, to give reasons for accepting them, and then for accepting their respective conclusion. In this example, however, nothing similar happens.

This allows us to understand a crucial limitation of arguing for the opposite in rational discussion: if we advance an argument whose conclusion is the negation of the claim initially proposed, we (1) do not give reasons to believe the premises of the counter-argument rather than those of the initial argument, and (2) we do not necessarily get a better understanding of the reasons supporting the initial claim. Let us go back to our example to see this.

The counter-argument in question introduces the premises P1 – P4, and from these the negation of the initial claim follows. If we accept P1 – P4, we cannot therefore accept the conclusion of the argument initially proposed. But why should we accept P1 – P4? We are not given any reason for this. We might well already accept these premises, but some of the audience may not. By advancing P1 – P4 we are just setting two arguments in opposition, where only one of the two can be sound. However, we are not giving reasons in favor of the premises of the counter-argument, and therefore the discussion is reduced to an arbitrary choice between its premises and those of the initial argument. If I choose the former, I will accept the conclusion of the comment; if I choose the latter, I will accept the conclusion of the post. However, I am not given any basis for choosing. Implicitly, the counter-argument in question has in fact doubled our task of ascertaining the acceptability of the reasons proposed, because the reasons expressed by P1 – P4 have no direct relationship with those expressed by the initial argument, and therefore proposing the latter does not offer us a means of evaluating the former.

This allows us to move on to the second point: the premises of the initial argument focused on forms of discrimination and their unacceptability. In principle, we might ask: is it true that any form of discrimination is unacceptable? Is it true that prohibiting same-sex marriage is discriminatory? The answers to these questions suggest whether the posted argument is sound (or not), and why. If, on the other hand, we limit ourselves to expressing reasons disconnected from these, and in favor of the denial of the initial claim, we will not have taken a single step towards understanding the strength of the reasons supporting the initial claim. In other words, the commentator is not considering whether prohibiting same-sex marriage is actually a form of discrimination but focuses on the alleged essential characteristics of the family. The two interlocutors hold opposing claims but are not comparing their own reasons. This is a problem, because the rational acceptability of a conclusion depends on the strength of the reasons expressed by the premises.

Graham calls the top of the pyramid (level 5) "refuting the central point"; here we call it "refutation". Let us imagine a last comment on the post: "Many, like you, think that banning same-sex marriage leads to discrimination, because homosexuals are treated differently from heterosexuals. However, this can be avoided by guaranteeing homosexual couples the same rights reserved for heterosexual couples". Let us extract the argument contained in this comment:

(P1) If we guarantee homosexual couples the same rights reserved for heterosexual couples, then there is no discrimination.
(P2) It is possible to guarantee both types of couples the same treatment, even without extending the institution of marriage to homosexual couples.

(C) It is possible not to extend marriage to same-sex couples without at the same time discriminating against them.

In this case, the comment is focused on the standpoint raised by the initial post, namely whether the prohibition of marriage between persons of the same sex *in itself* has discriminatory effects, and the counter-argument concludes that the prohibition of marriage between persons of the same sex does not, in itself, amount to a form of discrimination – although it does not exclude that discrimination will exist in the absence of some measures guaranteeing same-sex couples the same rights as married couples. Let us look briefly at the dialectical strategy of this argument, and at its logical relationships with the argument of the initial post. The conclusion of the argument, if true, implies that there is (at least) one logically possible scenario in which same-sex couples cannot marry but in which there is no discrimination (more precisely, the scenario in which the same legal rights are guaranteed to heterosexual and homosexual couples).

The conclusion of the proposed argument denies that the exclusion from marriage of same-sex couples is in itself discriminatory, and this implies that whoever proposes the initial argument must try to defend that premise, or qualify the premises and the conclusion (perhaps specifying the conditions under which the exclusion from marriage is discriminatory).

Interestingly, for Graham, in climbing up the pyramid of disagreement the discussions get better. The higher the levels of reaction to disagreement, the more rationally the different positions are tested. In addition, opportunities to consider aspects of the problem that have not yet been explored increase and, more generally, one's understanding of the problem is improved. Finally, the rational ways of reacting to disagreement favor the recognition and respect of others' positions, improving the practical effect of the discussion. This happens even though the discussion does not end with the acceptance of one of the various positions at stake. It may happen, in fact, that at the end of a discussion (even a perfectly rational one), everyone remains entrenched in their own position. Even if this happens, however, a genuine argument allows an intellectual exchange which enriches the interlocutors. This is impossible to achieve at the lower levels of the pyramid of disagreement.

3.4 Conclusion

Let us briefly summarize the content of this chapter. We have given a definition of rational discussion and we have looked at some of its "canonical properties" according to van Eemeren and Grootendorst. We have related these properties to the purpose of rational discussion: to contribute to an adequate and increasingly improved understanding of the claim expressed by an argument and of the reasons presented in its support (§ 3.2). We then discussed the possible ways of expressing disagreement, organizing them on a scale that goes from the least to the most rational (the so-called pyramid of disagreement devised by Graham), where the rationality of the kinds of expression of disagreement is evaluated on the basis of their potential to contribute to a rational discussion (§ 3.3).

In the next chapter, we will dwell in more detail on the last mode of expressing disagreement identified by Graham, namely refutation. We will look specifically at some possible strategies of attacking an argument in a rational way in order to prove that it is not good.

4 | How to reply rationally to an argument

As we saw in the previous chapter, in a rational discussion we should not attack our interlocutors nor claims that they did not make. One is supposed to attack the argument of others. But what is the best way to rationally counter-argue against the claim or some aspect of the reasoning of others? In this chapter, we will deal with this issue.

Learning to rebut arguments we do not accept is an important aspect of argumentation and a fundamental skill in Critical Thinking. After all, it is not difficult to realize that real discussions, whether they are conducted at a seminar for graduate students, in a brawling condominium meeting or in a courtroom, usually involve a dialectical exchange of reasons – more prosaically, a back-and-forth discussion.

A good argumentative contribution must produce refutations (or, at least, attempts at refutation) of the arguments we do not accept. This requires us to analyze the structure of the counterparty's reasoning, an activity that often leads us to grasp aspects of it that are not in full light and deserve to be known.

In this chapter we discuss the two main strategies we can use to rationally argue back (§ 4.1). The first one is to attack its argumentative structure (§ 4.2), that is, to point out that the premises do not provide reasons for the conclusion. The second strategy is to attack the truth of the premises, or their justifiability or plausibility (§ 4.3), that is, to show that there are solid reasons to doubt that the premises are true. We will discuss some epistemic aspects related to this strategy separately (§ 4.1.1). Both strategies presuppose the ability to correctly evaluate whether the argument initially proposed is a good argument.

4.1 The two strategies for attacking an argument

In Chapter 2 we claimed that a good argument includes true premises, which provide reasons to support the conclusion. This definition gives us a criterion to recognize bad arguments. To this purpose, we need to ascertain (i) whether at least one premise of the argument is not true, and (ii) whether the premises, jointly, do not provide reasons that underpin the conclusion. If at least one of the two conditions is met, then the argument under scrutiny is bad. Otherwise, we will recognize it as good. In short, the definition of a good argument provides general criteria for evaluating an argument. We can basically use the same criteria to show – or rather, argue! – that a given argument is bad. Once we have recognized it as such, we can articulate our counter-argument adopting one of the following two strategies:

I. show that the premises do not provide reasons to support the conclusion;
II. claim that at least one of the premises is false.

These strategies are strictly related to *two* evaluative dimensions of argumentation. The first one focuses exclusively on the logical or structural features of the argument. In the case of deductive reasoning, in this dimension we evaluate the validity (or invalidity) of the argument. In the case of non-deductive reasoning, we assess the strength (or weakness) of the argument. More generally, the logical or structural dimension is what we must look at when we ask ourselves: "Do the premises of this argument give reasons to support its conclusion?" When, in proposing counterarguments, we choose strategy (I), we focus on this evaluative dimension.

The second evaluative dimension is instead of an alethic-epistemological nature, that is, it focuses on the truth of the premises, or on their rational justifiability or plausibility.[1] We will see shortly why we are making reference to the notions of justifiability and plausibility alongside that of truth, which appears in our definition of a good argument. When, in replying to an argument, we choose strategy (II), we focus on this evaluative dimension and cast doubts on the truth of the premises, either directly, or by questioning their rational justifiability or plausibility.

The process of evaluating an inference takes place quickly and is, in a sense, instinctive. This is because some logical patterns are processed very easily by our brain – our cognitive and computational abilities allow us to manage them easily – or because we find premises whose truth (or falsity) is a shared belief.

4.1.1 Truth, justifiability and plausibility

Let us now consider more closely why, when outlining strategy (II), we mentioned not only the truth of the premises, but also their rational justifiability and plausibility. A sentence can be true or false, and this is the relevant fact with regard to condition (1) in the definition of a good argument (see § 2.3). Evaluating the truth of an argument's premise, however, brings another dimension into play: the epistemological one. This dimension is very important because in many cases we do not have access to the facts that make a sentence true or false. Consider the statement:

The notebook is in the kitchen and not in the bedroom.

Suppose this is true, but that I went out in the morning without checking where I left the notebook. If someone called me from home and claimed "The notebook is

---

1. The term "alethic" means "related to truth" and derives from the ancient Greek word "aletheia", which means "truth". Recall that "epistemological" (or "epistemic") means instead "related to knowledge". The notion of "rational justifiability" is primarily epistemological: it explains when a rational subject is justified in taking a sentence as true. Similarly, the notion of "rational plausibility" prescribes what a rational subject is justified in assuming to be true.

in the kitchen and not in the bedroom", I would not know whether this is true or false. But suppose that I always leave that notebook in the bedroom and did not realize that I had distractedly left it somewhere else. I would react by considering the above statement false, and would be ready to argue accordingly. Although I do not know for sure that it is false, I am nevertheless rationally justified in believing it to be false. Something similar occurs when one replies to an argument using strategy (II). Let us go back to the argument in the Facebook post:

(P1) Prohibiting same-sex marriage is a form of discrimination.
(P2) Any form of discrimination is unacceptable.
(C) Prohibiting same-sex marriage is unacceptable.

Suppose that P1 and P2 are true. Since the argument is deductively valid (and non-circular), it is a good argument. But suppose that a friend of ours believes that P1 is not true, and uses strategy (II) to respond to the argument. What does it mean that P1 is true with regard to this purpose? It implies that the respondent is attacking it incorrectly. However, if the respondent has solid reasons to believe that P1 is false, then she has the right to question its truth in the course of the discussion. The important thing is that her standpoint is rationally justified or rationally plausible. These are the two conditions under which the use of strategy (II) is legitimate, regardless of the truth of the premises. Similarly, we may not be certain that a premise is false, but if we have a rationally justified belief that it is so, or if we think it is rationally plausible that it is false, then we are entitled to put forward a counter-argument following strategy (II).

All this is of paramount importance because truth is not epistemically transparent. There are true (or false) sentences that we do not know to be such, or that we mistakenly believe to be false (or true), or about which we are uncertain. We are not free from error even when we use our cognitive and epistemic abilities as best as we can. Furthermore, there are many things that escape our knowledge. If we could only use strategy (II) when we were certain that one or more premises are false, then we would almost never be in a position to resort to it. This would be especially the case in situations where knowing the truth of the premises is more difficult – situations that are actually more interesting from the dialectical point of view, since they are more likely to generate disagreement between rational subjects.

At the same time, counter-arguments focused on the premises cannot be arbitrary and must be aligned with our main epistemic goal, namely that of "tracking the truth". This is precisely why we are entitled to criticize a premise even when we are uncertain about its truth, as long as we are justified in believing that that premise "does not track" the truth, or, more simply, when it is clearly not true. In a nutshell, the crucial point is the following: we are supposed to use strategy (II) when we have a rationally justified belief that a certain premise is not true, or is implausible. This explains why in discussing strategy (II) we have mentioned notions such as rational justifiability and rational plausibility, together with that of truth.

4.2 Attack on the argumentative structure

Imagine that a member of the Parmigiano Reggiano (PR) producers' association wants to convince the public of the superiority of her product over the Grana Padano (GP), a parmesan cheese produced in a different area of Italy. Suppose then that for this purpose she proposes the following argument:

(P1) GP is made with milk from cows fed with silage.[2]
(P2) It is proven that silages reduce the organoleptic properties of milk.

(C) GP is not a quality product.

Obviously, proving that GP is not a quality product is not the same as proving that PR is. It could well be that both GP and PR are poor quality products. However, we can think that the argument proposed above is only the first step of a complex argumentative strategy via which the member of the PR producers' association wants to show that PR is superior to GP. According to this strategy, she first makes it the case that GP is not a quality product, then she will show that her product is of good quality. How could one reply to this argument, or simply ascertain whether it is a good argument? As we have seen, we can attack its argumentative structure, or call into question the truth of the premises. Let us start with the first strategy. This is the more properly "logical" attack; in essence, it tries to show that the argument is either deductively invalid (in the case of deductive reasoning) or that it is a weak argument (in the case of non-deductive reasoning). Let us go back to the argument of the PR producer:

(P1) GP is made with milk from cows fed with silage.
(P2) It is proven that silages reduce the organoleptic properties of milk.

(C) GP is not a quality product.

Do P1 and P2 provide reasons for the truth of C? The answer is simple: they do not. P1 and P2 jointly provide reasons for concluding that GP has organoleptic properties inferior to the cheese produced by means of milk of animals fed with silage. Thus, it is not a dairy product of the best possible quality. However, this does not imply that it is not a quality product. To imply it, it is necessary to add one of these two premises:

(P3) Any dairy product that is not of the best possible quality is not a quality product.

or:

(P3') The inferior organoleptic properties of the milk by means of which GP is produced are such that they make GP qualitatively poor.

---

2. Silage is a feed resulting from the storage and fermentation of green or wet crops under anaerobic conditions.

Neither can be passed as an implicit premise. This is for two reasons. First, these premises do not constitute common knowledge or a general belief, at least if we assume that the discussion takes place in a public debate in which the participants lack industry expertise. Furthermore, the truth of these premises (assuming they are true) is not obvious and should not be taken for granted – indeed, it should be properly argued in case the public lacked specific skills.

More specifically, the argument of the member of the PR producers' association is such that P1 and P2 can be true without C also being true. Therefore, the argument, if we do not add further premises, is not deductively valid. It is a weak argument because the truth of the premises only modestly increases the probability that the conclusion is true. Therefore, it is a bad argument.

In the following paragraphs we will consider two very basic forms of fallacies. Remember that fallacies are bad reasoning patterns that real human cognitive subjects are inclined to use. We will take advantage of the first fallacy to see how we can respond to an attack against the logical structure of an argument. The game of replying rationally to an argument cannot consist of a single step. It develops through a succession of acts of reasoning in which the one who has received a reply adjusts the argument, or reacts to the reply of others by applying either strategy (I) or strategy (II). Each rebuttal, in short, can be followed by a rejoinder of some sort, and so on *ad infinitum*. After discussing the two types of fallacy, we will consider the only type of deductive argument that, even starting from true premises, cannot generate good arguments: circular reasoning, or vicious circle.

4.2.1 Non-sequitur

Consider the following argument:

(P) The milk used to produce GP comes from animals fed with silage.

(C) The milk used to produce GP is of low quality.

This is a very basic example of a *non-sequitur*. A *non-sequitur*[3] is, technically, any argument in which the conclusion does not necessarily follow from the premises – that is, any argument presented as deductively valid by the person who proposes it but whose conclusion does not necessarily follow from its premises. In this sense, labeling an argument as a *non-sequitur* is a canonical way of replying when the argument in question is supposed to provide conclusive reasons for the conclusion, even though this is not the case. Arguments based on a *non-sequitur cannot* be good arguments.

---

3. In the interest of simplicity, we do not discuss the non-deductive forms of *non-sequitur*.

This example also shows how one might respond to a rebuttal. Suppose that Eleanor proposes the argument mentioned above, and that Mary replies: "Yours is not a good argument, because it is a *non-sequitur*". Eleanor could argue that she left a premise implicit because she thought it was common knowledge. Eleanor could then reformulate her argument adding a further premise:

(P1) The milk used to produce GP comes from animals fed with silage.
(P2) A silage-based diet makes the milk of low quality.

(C) The milk used to produce GP is of low quality.

This argument is deductively valid. In fact, P2 asserts that a silage-based diet is sufficient for cows to produce low quality milk (notice that the argument discussed in the previous paragraph had a weaker premise, namely that silage simply reduces the quality of cheese). Coupled with the premise that Eleanor has already proposed in her initial argument, this implies the conclusion that the milk used to produce GP is of low quality.

What happens next in the rational argument? Did Eleanor close the game with the second version of her argument? Of course she did not. Mary can no longer claim that the argument is a *non-sequitur*, but she has something else to fall back on: she can focus on the additional premise P2 proposed by Eleanor. Faced with this premise, Mary can resort to strategy (II) and question its truth. How could that be done? If Mary has a rationally justified and soundly motivated belief that P2 is false, then she can argue for it being false. In doing so, she casts doubt on the fact that P2 is rationally justified or plausible, thereby indirectly attacking its truth. Finally, if Mary does not have specific expertise in dairy products (and does not know whether P2 is true or false) she is still entitled to ask Eleanor to justify P2. This is an indirect way of attacking the premise. Of course, Eleanor could have a very solid justification for P2, unbeknown to Mary, and this would give Eleanor a chance to counter Mary's further objection.

4.2.2 A fallacy of inductive reasoning: the unwarranted generalization

Fallacies plague not only deductive reasoning but also other types of reasoning. Inductive reasoning, for example, is exposed to one of the most basic and resilient fallacies in the context of our cognitive activities, namely the unwarranted generalization. Consider the following argument:

(P1) Dioxin was found in the milk of the Borgoforte dairy.
(P2) The Borgoforte dairy produces GP.

(C) GP contains dioxin.

As with many inductive arguments, the argument in question projects a property ("containing dioxin") from the totality of an observed sample (the GP produced by a given company) to the totality of a set of objects (all GP). What is wrong with this argument, then? The answer is simple: the fact that the observed sample is not sufficient to operate the projection. In the argument we are examining, this implies that P1 and P2 do jointly not increase the probability that the conclusion is true. Thus, the argument is not good because its premises give no reason to believe that the conclusion is true, even if they are true.

The unwarranted generalization – which we will discuss more at length in Chapter 9 – is, for inductive reasoning, what *non-sequitur* is for deductive reasoning. Indeed, the canonical way of replying to an inductive argument is to point out that the premises do not increase the probability of the truth of the conclusion, just as replying to a piece of deductive reasoning is to point out that the conclusion does not necessarily follow from the premises.

Another case of unwarranted generalization is the following:

(P1)   Parmigiano Reggiano is a quality product.
(P2)   Parmigiano Reggiano is an Italian product.

(C)   All Italian products are of quality.

This is a case similar to the previous one. The only relevant difference is that we could have further reasons to believe the conclusion to be true, in the event that we do appreciate Italian products in general. It is important to remember, however, that in the evaluation of an argument (and, therefore, when one tries to reply to it) what matters is not simply the truth of the conclusion, of which perhaps we can be certain on the basis of other considerations, but rather the way in which the conclusion is obtained starting from the premises. Consider another example:

(P1)   This raven lives in the Tower of London.
(P2)   This raven is black.

(C)   All the ravens living in the Tower of London are black.

Although the conclusion and the premises are true, this is not a good argument. The fact that a single raven living in the London Tower is black does not *per se* increase the probability of the conclusion that all the ravens living in the London Tower are black. This argument, therefore, is not good.

4.2.3 Circular reasoning

To understand how circular reasoning works, let us start with a very simple example. Take the following argument:

(P)   Parmigiano Reggiano is a product known all over the world.

(C)   Parmigiano Reggiano is a product known all over the world.

This argument is *deductively valid*: it is easy to see that if the premise is true, then the conclusion *must* be true. After all, they are the same sentence. However, this argument cannot be a good argument, even if P – and therefore C, since they are the same sentence – is true. In fact, *repeating* the same sentence as a premise and then as a conclusion gives no reason to support the sentence in question – at the end of the day, no reason is provided to support it.

The type of deductive reasoning in which:

(i) the conclusion of the argument also serves as a premise
(ii) no other premise provides reasons to support the conclusion

is called *circular reasoning* or *vicious circle*. Any form of circular reasoning is a deductive argument, because the conclusion also figures as one of the premises, and this warrants that if all the premises are true, the conclusion must also be.

With the simple example just proposed we have shown that there are deductively valid arguments that cannot be *good* arguments even when their premises are true. These are precisely the cases of circular reasoning. With the exception of the latter, however, deductively valid arguments that have true premises are good arguments. Consider the following example:

(P1)   Eleanor is eating Parmigiano Reggiano but doesn't know what she is eating.
(P2)   Mary is eating Parmigiano Reggiano but doesn't know what she is eating.
...
(Pn-1) Subject n-1 is eating Parmigiano Reggiano but doesn't know what she is eating.
(Pn)   Parmigiano Reggiano is a product known all over the world.

(C)   Parmigiano Reggiano is a product known all over the world.

This too is a circular reasoning. Premises P1, ..., Pn-1 clearly do not provide any reason in support of the conclusion C. The truth of the conclusion is trivially warranted by the truth of Pn (*if* the latter is true, of course), because Pn and C are the same sentence. Instead, the argument

(P1)   All human beings are mortal.
(P2)   Socrates is mortal.
(P3)   Socrates is a human being.

(C)   Socrates is mortal.

is *not circular*, although the conclusion coincides with one of the premises. In fact, if

P1 and P3 are true, C must be true. To put it differently, P1 and P3, if true, make the conclusion true, and also provide reasons to support it. The repetition of the conclusion in the premises is redundant here: the argument would do its job perfectly without that repetition. This allows us to present a general method for recognizing whether an argument is a case of circular reasoning. Given an argument of the form

$$(P1)$$
$$(P2)$$
$$\dots$$
$$(Pn)$$
$$\overline{}$$
$$(C)$$

if Pi is equal to C, being Pi a premise between P1 and Pn, we can drop Pi and construct the following argument:

$$(P1)$$
$$(P2)$$
$$\dots$$
$$(Pi\text{-}1)$$
$$(Pi\text{+}1)$$
$$\dots$$
$$(Pn)$$
$$\overline{}$$
$$(C).$$

If the premises of this new argument *do not* provide reasons to support C, then the initial argument is an instance of a circular reasoning. If, on the other hand, the premises provide reasons to support C, the initial argument is *not* an instance of circular reasoning. And if its premises are true, then it will also be a good argument. If more than one premise is equal to the conclusion, we subtract all those premises, and construct a new argument in which none of the remaining premises coincide with the conclusion. We then proceed as before, checking whether the new argument provides reasons supporting C.

We can extend the definition of a circular argument to other cases in which, although the conclusion does not appear among the premises, it is nevertheless presupposed by at least one of them. Suppose someone argues like this:

> God exists because the Bible says so and the Bible says true things because it was written by God and God is infallible.

In this argument the conclusion ("God exists") does not appear explicitly in any of its premises and yet these *presuppose* the conclusion. That God wrote the Bible presupposes the existence of God, which is also presupposed, at first sight, by the claim that God is infallible. In other words, the conclusion must be true for the premises to be true. Of course, it is quite obvious in this example that the premises presuppose the conclusion. In very complex arguments, however, this might be not apparent at all. Although no one will be persuaded by an argument in which the conclusion also serves as a premise, or is otherwise presupposed by one or more of them, arguments in which the conclusion is subtly presupposed by one of the premises can easily be misleading.

4.3 Attack on the premises

We have seen that a second strategy for replying to an argument is to attack the truth or rational plausibility of its premises. If we are able to show that one of the premises is false, then we prove that the argument is not good. In this case we directly attack the truth of the premises. The number of situations in which that can be done is rather limited and these are quite simple scenarios. Let us go back to the notebook example, and suppose the person who lives with me says:

The notebook is in the kitchen and not in the bedroom.

To show that this statement is false, all I have to do is open the door of my bedroom and show that the notebook in question is there. This is sufficiently conclusive proof of the fact; sufficiently conclusive because it is logically possible that we might have hallucinations at the exact moment we open the door, even if, unless specific conditions hold true, the chances of this happening are very low. This is a case in which the empirical evidence is so clear and elementary that it shows by itself, so to say, that a statement is false. There are also cases in which a sentence can be proved to be false because it is inconsistent – we will return to this shortly, when we will talk about the *reductio ad absurdum*.

However, empirical evidence is not always that clear – and above all, it is not always accessible to us and to our interlocutor. Furthermore, the sentences that we believe to be false are not always inconsistent. When we cannot directly show that a premise is false, however, we can act *indirectly* by questioning its rational plausibility. In this case we give reasons to doubt that the argument is good, because if its premises are not rationally justifiable, then we are entitled not to believe that they are true, and to reject them. This is also crucial in relation to the rules of rational discussion in § 3.2. At this point, rule 5 comes into play: whoever proposes an argument cannot take the truth of the premises for granted, unless their truth has already been proven or constitutes common knowledge.

There are numerous avenues that can be taken to refute a premise. In the following we will consider three of them: counterexamples, request for justification, and *reductio ad absurdum*.

4.3.1 Counterexamples

The method of counterexample is used in those cases where one of the premises of the argument we intend to criticize is a universal sentence. A sentence is universal if it states that all objects in a certain group relevant to the argument have a certain property. For example, "All swans are white" is a universal sentence.[4] The existence of at least one object (of the relevant group) that does not have this property is sufficient to falsify a universal sentence. For example, a single black swan falsifies "All swans are white". Any object that helps falsify a universal sentence in the way just outlined is called a "counterexample". When a premise we wish to question is a universal sentence, we can show it to be false if we are able to exhibit a counterexample, or if we can somehow prove or argue that it exists.

Let us return to the argument proposed by the member of the Parmigiano Reggiano producers' association. Its premises, even if they are not explicitly formulated as universal sentences, turn out to be such upon closer scrutiny. So we can reformulate that argument making its universal premises explicit:

(P1) All GP is made with milk from cows fed with silage.
(P2) There is evidence that silage feeding reduces the organoleptic properties of milk.

(C) All GP is not a quality product.

The first premise could be falsified if one is able to show that not all GP is made with milk coming from cows fed with silage; the second premise could be falsified if one is able to show that there are cases in which silage feeding does not reduce the organoleptic properties of milk. After all, silage feeding could be not that bad as far as the quality of the milk is concerned.

4.3.2 Request for justification

A weaker way to attack a premise is to require a justification of its truth, a move we have already discussed in § 4.2.1. Here we add other considerations that help us understand the general role of this option in responding to an argument. The burden of proof falls on the person proposing an argument, as rules 3 and 5 of Rational Argumentation prescribe. The reason is clear. The subject who advances an argument should, if asked, provide a justification of the premises of the argument, unless they make universally accepted claims (in which case, the burden of proof is on the shoulder of the subject who does not accept the premises). If the person proposing an argument is unable to justify the premises they have used, then we have good

---

4. We will discuss this type of sentence at greater length in the next chapter.

reason not to accept those premises. Such premises may well be true but in the absence of adequate justification by the proposer we have no rational basis for accepting them, should we not already agree with them. Consequently, we have good reason to think that the proposed argument is not good. It is important, in fact, that those who propose an argument use not only true premises but premises that are rationally acceptable to the other participants in the discussion. Rationally justifying a premise is the best way to ensure its acceptance.

We have seen that asking those proposing an argument to justify their premises can be strategically useful. If the interlocutor fails to provide adequate justification, then it leaves the way open for a critical reply.

4.3.3 Reductio ad absurdum

Reductio ad absurdum (RAA) involves refuting an argument by showing that one of its premises results in a contradiction. This is a method of demonstration that is mainly applied in mathematical reasoning. Suppose we assume A, and that from A, along with other statements that are part of our mathematical background knowledge, we derive a contradiction, namely a sentence of the form "B and not B", which cannot be true. We can validly conclude "not A" from it – that is, we can conclude that A is not true, because a sentence that implies something false cannot be true (we will discuss this principle in Chapter 6).

Let us illustrate this with an elementary example. Suppose I say, (0) "There exists a natural number i that is larger than all other natural numbers". It is a basic assumption of mathematics that (1) for every natural number n, there exists the natural number $n + 1$; (2) for every natural number n, $n + 1$ is larger than n. By (1)-(2) we show that, given the number i of premise (0), there exists the number $i + 1$ and that such a number is larger than i. From (0)-(2) together we derive the conjunction "There exists a natural number i larger than all other natural numbers, and there exists a natural number $i + 1$". This is, of course, a contradiction. Since this cannot be true, and (1)-(2) follow from some basic definitions about the natural numbers, having derived a contradiction is sufficient for us to conclude the negation of (0): there is no natural number i larger than all other natural numbers.

In everyday life we often use a form of reasoning similar to the RAA but less rigorous. Specifically, in some cases we express disagreement with an argument because one of its premises seems absurd, implausible, or unacceptable. Suppose someone presents an argument which includes the following premise: "It is right to liberalize all drugs, even so-called hard drugs". One might reply that this premise is unacceptable because the financial burden of addiction treatment and rehabilitation programs would grow disproportionately and become unsustainable.

This improper form of *reduction ad absurdum* is rather tricky. In general, it is difficult to prove that a conclusion is "unacceptable" unless it is logically unacceptable (i.e. a contradiction). Even in cases where a good counter-argument can be presented, as in the example of drug liberalization, it is almost never trivial to prove that the conclusion of the initial argument is unacceptable. Are we sure that the economic issue of health care spending should be given greater weight than individual freedom? Is it not true that the limitation of individual freedom is actually unacceptable? Shouldn't we rather modify government spending by cutting elsewhere in order to provide health care where that freedom is exercised? Couldn't it be that the maintenance of a certain distribution of spending is unacceptable? Who is right and who is wrong here?

Even leaving aside the practical acceptability of some decisions, in general the fact that something seems unacceptable to us is not a conclusive indicator that it is false, even if we believe we have good arguments against it. After all, mankind believed until five hundred years ago that it was the Sun that revolved around the Earth, and as absurd as it seemed to the contrary in light of pre-Galilean physical theories, this belief was false.

4.4 Conclusion

In this chapter, we have focused on the most rational way of expressing disagreement, which allows us to respond to an argument in a fully rational manner. This can be implemented through two distinct strategies, which can be combined together. The first strategy (§4.2) consists of attacking the argumentative structure. It relies on the fact that the premises of the argument to which one is replying do not actually give reasons to support the conclusion. The second strategy (§ 4.3) calls into question, directly or indirectly, the truth of the premises. It is used when we have solid reasons to believe that the premises of an argument are false. The argumentative structure can be attacked in many cases; the ones we have discussed concern the "canonical fallacies" of deductive reasoning (§ 4.2.1) and inductive reasoning (§ 4.2.2), and the circularity of an argument (§ 4.2.3). We have seen that circular reasoning is the only form of deductively valid reasoning that cannot yield a good argument. The truth of premises, too, can be attacked in many ways. The ones we have considered in this chapter are the appeal to counterexamples (§ 4.3.1), and the demand for a justification (§ 4.3.2). In addition, there is a purely logical method of showing that one or more premises are false: showing that they imply a contradiction (*reductio ad absurdum*, § 4.3.3).

Part II
Deductive arguments

5 | Deductive arguments

In Chapter 2, we discussed the distinguishing feature of deductively valid arguments: they secure the truth of the conclusion, given the truth of the premises – that is: it is *logically impossible* for the premises to be true and the conclusion to be false. Or, in other words: asserting the premises but denying the conclusion means running into a contradiction. In this chapter we will analyze deductive arguments in greater detail and we will try and identify rigorous criteria that help us distinguish deductively valid arguments from arguments that are not deductively valid.

This is of crucial importance because deductive reasoning plays a role in all human intellectual activities. We use it in any area of knowledge – Economics, Law, Botany, Zoology, Philosophy, Astronomy, and so on. Besides, we use it every day. For example, from the fact that Anna is either at home or at the gym and the fact that she is not at home, we can *deduce* that she is at the gym. These are two of the many reasons why it is crucial to distinguish those cases of deductive reasoning that are *valid* (and can thus actually perform their function in our reasoning) from those that are not valid (and therefore fail to play the role that a deductive argument carries out in our reasoning). Deductively valid arguments come in a wide variety of forms. Here we will exemplify some of the main ones, in order to provide the reader with a basic view of some of the forms of deductively valid reasoning that we encounter most frequently in our reasoning activity.

5.1 Introduction

Let us go back to one of the deductive arguments we mentioned in Chapter 2:

(P1) All sculptures imitate natural objects.
(P2) *Bird in space* does not imitate a natural object.

(C) *Bird in space* is not a sculpture.

We have mentioned already that this argument is deductively valid, and it is easy to see that, indeed, *if* its premises are true, then its conclusion is *also* true. However, if this argument is valid, the following one will also be valid:

(P1) All sculptures imitate natural objects.
(P2) *Unique forms of continuity in space*[1] does not imitate a natural object.

(C) *Unique forms of continuity in space* is not a sculpture.

Exactly as the next one:

(P1) All horses have four legs.
(P2) Furia does not have four legs.

(C) Furia is not a horse.

Obviously, we can realize that for these two arguments too, there is no possible scenario in which the premises are true, and the conclusion is false. There is a very important reason why this is the case for all three arguments in question: they all share *the same argumentation scheme*.[2] It is precisely due to this kind of scheme that, in each of the three arguments, the two premises cannot be true without the conclusion also being true.

In other words: if we wish *to understand* whether an argument is deductively valid or not, we build logically possible scenarios in which the premises are true, as we said in Chapter 2. More specifically, if we can build at least one scenario where the premises are true and yet the conclusion is *false*, then we show that the argument is *invalid*; by contrast, if every scenario where the premises are true is such that the conclusion is also *true*, then the argument is valid. However, whether or not there are possible scenarios that invalidate the argument is not due to chance or contingent facts: it is due to its argumentation scheme. Thus, whether an argument is deductively valid or not ultimately depends on its argumentation scheme.

What exactly *is* an argumentation scheme? Let us recall an important point first: in Chapter 2 we also said that the validity of an argument does not depend *on its relations with the world* but *on the relations between the concepts involved in the premises and in the conclusion*. And it is easy to see that, in fact, the three arguments above propose connections (or relations) between the concepts they mention, and these connections (or relations) share the same structure,[3] namely: in all three cases the premise (P1) "puts" all the elements of a set among the elements of another one (all the sculptures among the imitations of natural objects, all the horses among the animals that have four legs). The premise (P2) states that a particular individual (*Bird in space*, *Unique forms of continuity in space*, Furia) does not belong to the wider set. The conclusion (C) says that therefore that individual does not even be-

---

1. It is a sculpture by Umberto Boccioni. There are different versions of it by the artist, and they all have the same name.
2. We can express this in an equivalent way by saying that all these arguments share the same *logical form*.
3. Indeed, that is why we say that they have the same logical form.

long to the smaller set because, if it is not in the wider one, then it is not in the smaller one. We will return to the argumentative structure of the three examples in § 5.3. For the time being, notice that the validity of an argument depends on the relationships between the *logical* concepts involved in the premises and in the conclusions. What does this mean? An example will make this easy to understand. Take the following elementary case of reasoning:

(P1) Riccardo is driving and is not answering a call on his cell phone.
───
(C)   Riccardo is driving.

This reasoning is *valid*: if P1 is true, so must C. This, however, is not due to what P1 and C are about – that is, the facts expressed by "Riccardo is driving" and "Riccardo is not answering a call on his cell phone". Instead, it is due to the conjunction "and" which occurs in P1). A conjunction is what logicians call a *connective* or a *sentential operator*: a syntactic element that applies to one or more sentences (two, in the case of the conjunction in P1) and give us a further sentence in return. For example, if I apply the conjunction to sentences A and B in this order, I get the sentence "A and B". With some ambiguity, we also call "conjunction" any sentence of the form "A and B" – that is any sentence that results from the application of the connective to two "smaller" sentences. There is no risk of confusion here: the context will make it clear whether we are talking about the connective or the sentence. Otherwise, we will specify the point explicitly.

The conjunction (connective) *works in such a way* that from "A and B" we can always validly infer any of A or B. We can give an explanation of this in terms *of truth-values*: "A and B" is *true* if and only if both A and B are *true*. There is therefore no scenario in which any conjunction (sentence) "A and B" is true, and A is not true (or B is not true). Hence, there is no logically possible scenario in which "A and B" is true, and A is false. If we replace A with "Riccardo is driving" and B with "Riccardo is not answering a call on his cell phone", we have the explanation of why the reasoning that led us from P1 to C is deductively valid: it is the presence of the conjunction as a "main connective" of P1 to make it so – we will see in § 5.2 what the "main connective of a sentence" is.

As for the three initial arguments, we will see in § 5.3 that the logical concepts involved there – and which secure the validity of the three arguments – are that of the universal quantifier "All" (or "For every"), and that of the conditional connective "If ... then". The presence of the latter may sound unexpected, but we will see in § 5.3 why it is actually involved.

Let us use the example with the conjunction in order to understand what an inference scheme is. In "going from" the premise "Riccardo is driving and is not answering a call on his cell phone" to the conclusion "Riccardo is driving", we have made an inference that involves two specific sentences: the conjunction "Riccardo is driving and is not answering a call on his cell phone", and the first of its conjuncts. But of course, we are ready to acknowledge that the inference from "Roberto is writing and his little daughter wants to play with him" to "Roberto is writing" is the same *kind* of inference, because we do the same thing with *other* sentences: indeed, we start with a conjunction and conclude one of its conjuncts. This suffices to show that the nature (or identity) of an inference scheme does not depend on the particular identity of the sentences involved in a particular case of the scheme (say, on the fact that the premise tells us that "Riccardo is driving and is not answering a call on his cell phone" rather than "Roberto is writing and his little daughter wants to play with him"). The identity of the scheme, indeed, depends on the logical concepts *involved* in the sentences. The inference scheme of our example can in fact be "read" in a very general, abstract way as: "Assuming *any* conjunction, you can conclude any of its conjuncts".

The example of the conjunction also helps us notice a very important thing: the sentences "Riccardo is driving" and "Riccardo is not answering a call on his cell phone" do not make a real contribution to the validity of the argument. In other words: we could replace them with any pair conjunction/conjunct and the argument would remain valid, *as long as we leave the (connective) conjunction in its place*. Precisely for this reason, in this volume we will use not just specific sentences, but also *schematic letters*, or letters that, ideally, can stand for *any* sentence. In this volume we will need schematic letters for *sentences, predicates, individual constants* (ideally, *names*) and *individual variables*. In particular, we will use:

- $p, q, r, s, \ldots$ for a particular kind of sentences, which we will call *atomic sentences* (see § 5.2 for this);
- $A, B, C, D, \ldots$ for another kind of sentences, which we will call *arbitrary formulas*; they include both atomic sentences and the so-called complex sentences (see again § 5.2);
- $P, Q, R, \ldots$ for predicates (we will tend to use R for relational predicates, i.e. those that have two or more places);
- $a, b, c, \ldots$ for the individual constants (names);
- $x, y, z, \ldots$ for individual variables whose possible values are individuals (whether they have a name or not).

The chapter proceeds as follows. In the remainder of the Introduction (§ 5.1.1) we will briefly present the notions of *semantics* and *syntax*, and we will mention *classical logic, which is the system of logic that we actually follow in this volume when talking about deductive reasoning*. In § 5.2 we will endow ourselves with the tools for a rigorous understanding of propositional reasoning, that is, that reasoning in which just those logical operators that we call "connectives" are involved (in our case: negation, conjunction, disjunction, conditional, biconditional[4]). To do this,

---

4. We will postpone discussion of conditionals and biconditionals to Chapter 6.

we will present the basics of "propositional logic". We will see how its language is built up, distinguishing between two different kinds of sentences (simple and complex sentences); we will introduce the truth conditions for complex sentences (which depend on which is the so-called "main connective" within them) in § 5.2.1 and 5.2.2, and we will also see how they are crucial to establishing whether an argument (inference, reasoning) is valid (relative to arguments of propositional reasoning) in a concrete and rigorous way. Subsequently, in § 5.2.3, we will introduce a package of inference rules known as *introduction rules* and *elimination rules* (which together form the basis of *natural deduction* for propositional logic). In § 5.2.4 we will introduce a reading of these rules that helps us adjust them to contexts of reasoning in which the standards of our argumentative procedures are somewhat looser than the standards of mathematical proof – indeed, the rules in question were originally introduced in order to formalize mathematical reasoning.[5] In § 5.3 we will move on to more complex forms of reasoning, which involve *predicates* and *quantifiers*. In order to do this, we need what is called "quantified logic", or "predicate logic". We will build the relevant language (§ 5.3.2) and introduce the truth conditions for the new kinds of sentences. We will then use these conditions to check the validity or invalidity of some inference schemes, including the *syllogistic* inference schemes (§ 5.3.4). In § 5.3.3 we will then comment on some aspects of the formal regimentation introduced in § 5.3.2.

5.1.1 Semantics and syntax. Classical logic

In logic we can approach our reasoning from two different angles. The first "angle" considers deductive arguments based on the notion of validity that we have already introduced in Chapter 2, and it goes under the name of *semantics*. The second "angle" considers the arguments on the ground of their conformity to a set of rules of inference, which are all, ideally, derivable from a smaller "primitive" set – that is, a set of rules chosen for precise reasons as the economic basis for deriving the other rules. This second "angle" goes under the name of *syntax*. Let us briefly give some general insight on these.

Semantics is the branch of formal logic that (i) specifies the truth conditions of the sentences, and especially of those sentences containing the so-called "logical words" – for instance, connectives or quantifiers. Specifying their truth conditions means telling under what circumstances a sentence is true, or else what conditions must occur for the sentence to be true. Semantics also aims at (ii) interpreting other logical symbols (such as names, predicates, and individual variables). It is the

---

5. As we have said, we will postpone discussion of introduction and elimination rules for the conditional to Chapter 6.

branch of logic that we implicitly call into question when we ask ourselves if an argument is deductively valid, or when we ask ourselves if a sentence such as "Roberto is driving and is not answering a call on his cell phone" is true. When we specify the truth conditions in propositional logic and verify some validity (§ 5.2.1), we are applying semantics.

Syntax is that branch of formal logic that aims at: (i) specifying the rules for the formation of sentences, independently of their possible interpretations – the goal is to specify which combinations of symbols *can* have an interpretation or a truth-value, and which ones cannot. Syntax also (ii) specifies the rules of inference that help to legitimately "go" from one set of sentences to another in reasoning – that is, *to make valid inferences*. When we specify the introduction and elimination rules for the connectives in propositional logic (§ 5.2.3), for instance, we apply syntax.

Thanks to the development of a rigorous mathematical approach to logic, we are now aware that we can define a *large number of different formal logical systems* that have a range of desirable properties (in terms of the rules of inference they propose, for example). Many of them can be used as reasoning tools because they secure families of inference rules that we deem intuitive, or that we deem rational to apply. In short, we live in an era of *logical pluralism*. In this volume, we will follow one of the many systems defined by logicians, the one that has been considered, from the mid-nineteenth century to the end of the twentieth century, *the* most suitable system for prescribing rules of inference. This is called *classical logic*.[6] From a syntactic point of view, we could say that classical logic is defined by the rules of introduction and elimination of connectives and quantifiers that we will look at in this chapter and in Chapter 6 (for the material conditional), plus a few axioms, that is, statements considered logical principles. From the semantic point of view, we could say that classical logic is fundamentally defined by the truth conditions of the connectives that we will see in this chapter and in Chapter 6 (as regards the material conditional). In any case, it is worth saying something more intuitive about classical logic. We will do this from a semantic perspective, noting that:

1. Classical logic assumes that sentences can be either true or false. It does not allow for "intermediate values" such as "indefinite", "indeterminate" or the like. More precisely: classical logic is, semantically speaking, *bivalent*; in other words, it assigns to each sentence either the value "true" (T) or the value "false" (F). T and F are therefore called *truth-values*.
2. The semantical definition of negation in classical logic is such that *either* a sentence A is true, *or* its negation ¬A is true. This is because the latter is true if

---

6. Classical logic has actually had a remarkable competitor since the 1930s, namely: intuitionistic logic. The two systems were the main competitors for the role of a reference normative system in the area of human reasoning – with a particular attention to mathematical reasoning.

and only if A is false, and it is false if and only if A is true. Consequently, the disjunction A ∨ ¬A ("A or non-A") is a *logical truth* (or rather, a "classical logical truth") – that is, it is true in every logically possible scenario (according to classical logic). Furthermore, what we have said above implies that, in classical logic, we can read A ∨ ¬A as "A is true or A is false". In other words, in classical logic that disjunction ends up expressing the fact that the semantics is bivalent.

3. The semantical definition of negation in classical logic is such that a sentence cannot be both true and false – or rather: it cannot be the case that *both* A *and* ¬A are true. Hence, the conjunction A ∧ ¬A ("A and non-A") is a *logical falsity* (or rather, a "classical logical falsity") – that is, it is false in every logically possible scenario (according to classical logic). Classical logic is said to be *consistent* because it excludes the possibility that contradictions are true.

Not all formal systems follow points (1) – (3). Some are built in the belief that we should account for some sentences being *neither true nor false*. There are two large families of logics that do this. Philosophers call them, respectively, *paracomplete logics* and *logics with gaps*. Paracomplete logics do not follow point (2): A ∨ ¬A is not a logical truth in such systems. There is also intuitionist logic, which rejects the logical truth of A ∨ ¬A on the grounds that a disjunction should be accepted only if we have *either* a procedure to prove the first disjunct, *or* a procedure to prove the second. This is not always the case, not even when the two disjuncts are A and its negation ¬A. Other logics reject point (3), and believe that our formal systems must allow us to reason even in the presence of contradictions. These logics are called *paraconsistent*. We therefore have many possible logics, and they all offer an illuminating perspective on reasoning.

The reason why we will follow classical logic in this volume, is that it is the simplest logic, both from a semantic and from a syntactic point of view.

5.2 Propositional reasoning

Let us start with those forms of reasoning in which *connectives* are the only logical concepts involved – remember that these are operators that apply to sentences to generate other sentences. The branch of logic that deals with this kind of reasoning is that called "propositional logic" or "sentential logic".

In propositional logic, *sentences* are the smallest syntactic units; in other words, propositional logic is indifferent to those syntactic elements that are "smaller" than a sentence, such as names ("individual constants") and predicates. Furthermore, for propositional logic, connectives are the only logical elements determining reasoning connections between different sentences. We will see shortly what this means exactly. For now, let us try to gain a more concrete understanding of what is being said.

Propositional logic traditionally distinguishes five connectives:

- Negation: not; symbol: ¬
- Conjunction: and; symbol: ∧
- Disjunction: or; symbol: ∨
- Conditional: if ... then; symbol: →
- Biconditional: if and only if; symbol: ↔

Negation is the only "one-place" (monadic) connective among these connectives, with the other ones being "two-place" (dyadic) connectives. In propositional logic, the two basic building blocks in the construction of sentences are the connectives and what are termed *atomic sentences*, namely: those sentences that contain no connective (or expressions such as "all", "some", and so on) and therefore cannot be broken down into other "smaller" sentences. To get an idea of what these sentences would look like in our natural language, take the sentences:

(1) Anna is turning the key.
(2) The car started.
(3) This summer we will go on holiday to the sea.
(4) Anna loves her neighbor.
(5) Paolo is sick.
(6) Paolo will come to the appointment.

As we can see, these sentences do not contain any of the five connectives we have listed. In order to understand properly how propositional logic defines atomic sentences, however, we must abstract from the syntactic structure *within the sentence*. In order to get the point, remember that, in a certain sense, propositional logic does not "recognize" nouns, verbs and adjectives – or rather: it ignores them because it deals with reasoning connections that are completely independent from these elements. Keep in mind that propositional logic uses the lowercase Latin letters $p, q, r, s, \ldots$ for atomic sentences, and that these letters are, together with connectives, the minimal building blocks of all propositional logic. That being said, natural language sentences such as the ones we have exemplified provide a great intuitive way of understanding atomic sentences, and we will often use similar natural language sentences in order to exemplify atomic sentences. That is, we will assume that the atomic sentences of propositional logic express precisely those sentences of natural language that do not contain connectives, such as the ones above.

Once we have our stock of connectives and atomic sentences, we can define all the *possible* sentences of propositional logic, based on what is called a *recursive* definition of the sentences (or "formulas", as we also say in logic):

1. Every atomic sentence is a sentence.
2. If A is a sentence, then so is ¬A.
3. If A, B are sentences, then so is A ∧ B.

4. If A, B are sentences, then so is A ∨ B.
5. If A, B are sentences, then so is A → B.
6. If A, B are sentences, then so is A ↔ B.
7. Nothing else is a sentence.

All the sentences that contain connectives and can therefore be analyzed into "smaller" sentences – or to be exact, "simpler" sentences – are called "complex sentences". Some examples:

(1)   I do *not* like sushi.
(2)   Anna turned the key *and* the car started up.
(3)   This summer we will go on vacation to the beach *or* we will go on a 15-day trip.
(4)   *If* Paul is sick, *then* he will not join us.

Notice that 1 is considered a complex sentence since it contains a logical connective, that is: negation. We speak of *arbitrary formulas* when it is indeterminate whether we are dealing with an atomic sentence or a complex one. In propositional logic, the Latin capital letters A, B, C, D, ... are used for such formulas. It should therefore be kept in mind that A does not necessarily have an internal logical complexity (that is: it may not contain connectives). We cannot then in principle exclude that it stands for an atomic sentence: the kind of symbol, in itself, does not tell us whether we are facing an atomic or a complex sentence. This will not be a source of trouble, however, because we will use letters for arbitrary formulas when what we say is independent of whether we are talking about atomic sentences or complex ones.

Please bear in mind that by "negation", "disjunction", "conjunction", "conditional", "biconditional" we will denote not only the connectives, but also the sentences whose main connectives are, respectively, negation, disjunction, conjunction, ... (we will see in a few lines what we exactly mean by "main connective"). Hence, we will say that 1 is a negation, 2 a conjunction, 3 a disjunction, etc. As we have already said, the context will clarify when "negation" ("conjunction", "disjunction", etc.) refers to a sentence or rather to a connective (when necessary, by contrast, we will specify it).

One last thing before we go further. Connectives can be "nested" in a sentence, thus creating sentences of varying complexity. Take for example "*If* Paul is sick, *then* he will *not* join us". In symbols:

$$p \rightarrow \neg q$$

There are two connectives here. One (negation ¬) is used to construct one of the sentences that make up the conditional $p \rightarrow \neg q$, that is $\neg q$. The other (the conditional →) serves to construct the conditional sentence in its entirety. In general, complex sentences can have as large a complexity as desired, but this complexity will always only include a *finite* number of connectives and sentences, down to a finite number of atomic sentences that cannot be further analyzed. In the next few pages, we will see many examples of complex sentences in which many connectives are nested, and so we do not dwell on other examples here. However, it is important to make a distinction between the main connective of a sentence, and connectives that play, in a sense, a secondary role in the sentence. In $p \to \neg q$, the *main* connective is $\to$, because the sentence in its entirety is the result of the application of *that* connective to the two immediately simpler sentences, that is p and $\neg q$. A secondary connective in $p \to \neg q$ is $\neg$, because it is present in the conditional, but it is not that connective that "builds up" the conditional. On the other hand, $\neg$ is the main connective of $\neg q$, because $\neg q$ is the result of applying $\neg$ to q, and the latter is immediately simpler than $\neg q$, and, together with connective $\neg$, it makes it up $\neg q$. Other cases can be more complex but identifying the kind of sentence that we are facing proves very helpful in general. Take, for example:

$$\neg A \land \neg B$$

The main connective here is $\land$, since the sentence before us is obtained by applying the operator $\land$ to the sentences $\neg A$ and $\neg B$. However, in

$$\neg (\neg A \land \neg B)$$

the main connective is $\neg$, because the sentence in question derives from the application of the operator $\neg$ to the sentence $\neg A \land \neg B$. In general, the key to identifying the main connective in a sentence is to understand from which "immediately simpler" sentences that sentence is built on, and by using which connective. The rules of inference that we will see in this chapter apply to sentences according to what their *main* connective is.

The last formula we have exemplified gives us the chance to notice that it is difficult to read $\neg (\neg A \land \neg B)$ as "Not (not A and not B)" and, above all, this reading does not sound grammatically correct. This is because the recursive definition of formulas helps us build sentences of a syntactic complexity that is greater than that we experience in natural language. We will solve this problem by reading $\neg (\neg A \land \neg B)$ as "It is not the case that (it is not the case that A and it is not the case that B)" in a sort of mixed reading between natural language and formal language. In general, we will read $\neg A$ as "it is not the case that A".

5.2.1 The truth conditions of complex sentences

The connectives that we have considered help make the analysis of the sentences containing them particularly clear and simple. The main reason is that these connectives are *truth-functional*, that is: the truth or falsity of the sentences that contain them depends *entirely* on the truth or falsity of the simpler sentences that make them up, according to a rule that varies from connective to connective. To give an example: a conjunction such as

Anna turned the key and the car started up

is true if and only if *both* conjuncts ("Anna turned the key", "The car started up") are true – and it is false if and only if *at least one* of the two conjuncts is false. Using a mathematical terminology, we can say that the truth-value of a conjunction, disjunction, negation, etc. is a *function* of the truth-values of the sentences that compose them (the conjuncts, the disjuncts, the negated sentence, etc.).

Each connective determines a different *truth function* – that is, a function that applies to a truth-value, or to a pair of truth-values, to return a truth-value as an *output*. All these different functions can be visually represented through the so-called *truth tables*. Here we will look at the truth tables of negation, conjunction and disjunction. For the truth table of conditionals and biconditionals, we refer readers to Chapter 6, which deals in detail with reasoning using indicative conditionals, and in particular with the *material* conditional, which has a truth-functional semantic base. Let us start from the simplest truth table: the table of *negation* (¬).

Let us take the two sentences:

I like sushi.
I do not like sushi.

The second is true if and only if the first is false. And since a sentence is *either* true *or* false, we can also take it that the second sentence is false if and only if the first is true. In short, negation flips the truth-value of the sentence to which it applies. According to this, the truth table of ¬ is as follows:

A	¬A
T	F
F	T

In the table, T stands for "true" and F stands for "false". The table reads like this: the first column gives us the possible truth-values of A, the second the truth-values of ¬A, in relation to the given values of A; the rows, then, tell us how the given value of A determines the value of ¬A. The table that we have just depicted expresses the following instruction:

- With A – True: ¬A – False
- With A – False: ¬A – True

It is easy to see from the truth table or from the instruction above that ¬ is a truth-functional connective: all we need, in order to determine the truth-value of

¬A, is the truth-value of A, according to the rule imposed by the truth table of ¬. The tables also make it clear in which sense exactly a certain *truth function* is associated with connective ¬: the connective determines a function that, when taking T as an input, gives F as an output, and when it takes F, it gives T.

Let us go to *conjunction* (∧). We have already said that a sentence like

Anna turned the key and the car started up

is true if and only if both conjuncts are true, and it is false if and only if at least one of them is false. The following truth table summarizes this rule:

A	B	A ∧ B
T	T	T
T	F	F
F	T	F
F	F	F

The table is to be read as follows: the first two columns provide possible truth-values to A and to B. Four rows appear because we have four possible combinations of the truth-values of the two sentences. The last column provides the values of A ∧ B, in relation to the given values of A and B. Thus, the rows tell us how any combination of values of A and B determines a truth-value for A ∧ B. The table in question expresses the following instruction:

- With A – True and B – True: A ∧ B – True
- With A – True and B – False: A ∧ B – False
- With A – False and B – True: A ∧ B – False
- With A – False and B – False: A ∧ B – False

Also in this case, it is easy to see that we are dealing with a truth-functional connective: all we need in order to determine the truth-value of A ∧ B are the truth-value of A and the truth-value of B, according to the rule set by the truth table of ∧. Once again, it is clear that conjunction determines a truth function, and more precisely the two-place function which, if taking T and T as arguments, gives T, and gives F in all the other cases.

As for *disjunction* (∨), take the following sentence:

Either Paul is eating an apple or Paul is watching TV.

In order for this sentence to be true, it suffices that one of the two disjuncts is true. In other words, if it is true that Paul is eating an apple, then the disjunction is true, regardless of the truth-value of the other disjunct. Similarly, if Paul is watching TV, the disjunction is true, regardless of the truth-value of "Paul is eating an apple". By contrast, if *both* disjuncts are false, then the disjunction is false. The truth table of ∨ is the following:

A	B	A ∨ B
T	T	T
T	F	T
F	T	T
F	F	F

The above table expresses the following instruction:

- With A – True and B – True: A ∨ B – True
- With A – True and B – False: A ∨ B – True
- With A – False and B – True: A ∨ B – True
- With A – False and B – False: A ∨ B – False

Again, it is easy to see that ∨ is a truth-functional connective: in order to determine the truth-value of A ∨ B, all we need to establish are the truth-value of A and the truth-value of B, according to the rule set by the truth table of ∨. It is also clear that disjunction determines a truth function and, more precisely, the two-place function which gives T back as an output, if T is one of the two arguments, and gives F back only if no T appears as an argument.

As is clear from the first row in the table, ∨ expresses the so-called "inclusive disjunction": this means that this disjunction is true not only when only one of the two disjuncts is true, but also when both disjuncts are true. The disjunction in "Paul is eating an apple or Paul is watching TV" sounds inclusive, if only because the two disjuncts are compatible. But other disjunctions would sound "exclusive", that is, they would sound like they are false if both disjuncts are false or if both disjuncts are true. In natural language, we have no distinct syntactic elements for the two disjunctions, and it is often the context, emphasis, or other factors that tell us which one we are facing. If the two disjuncts are incompatible, then we are probably *using* the disjunction to express *exclusiveness* – for example, if we say, "Either Peter is a good man, or he isn't". In any case, we cannot say whether the disjunction is inclusive or exclusive simply by looking at the syntax of the sentence: in natural language "or" sometimes expresses an exclusive disjunction, and sometimes it expresses an inclusive disjunction. In logic, on the other hand, a distinct connective – xor – is used just for exclusive disjunction. The truth table of xor is the following:

A	B	A xor B
T	T	F
T	F	T
F	T	T
F	F	F

Let us return for a moment to the syntactic indistinguishability of the two disjunctions in our natural language. Take:

> Paolo is either at home or at Anna's.
>
> The subsidy is reserved for families with an income of less than *x* euro or with at least three children.

In the first sentence the disjunction sounds exclusive: it states that Paolo is at home or at Anna's but, obviously, he cannot possibly be in both places. In the second case, the disjunction sounds inclusive: obviously, in order to have the subsidy, one must be in one of two conditions – having a certain income or having at least three children – but it is clear that the subsidy will also be paid to those who fulfil both conditions. It has been argued that disjunction in natural language is actually always inclusive, and that the meaning component that excludes the truth of both disjuncts is actually implicitly conveyed. For example, we exclude that Paul is both in his house and in Anna's house not because this is told us by the use of "or", but because it is impossible for a human being to be in two different places at the same time. In this volume we will disregard these complications: we will follow the standard choice in logic and "identify" disjunction with inclusive disjunction ∨.

5.2.2 Definability

In propositional logic, many of the five connectives are *definable* starting from combinations of just some of them, and in classical logic (as in many other logics, actually) we can define all connectives on the basis of only a pair of connectives.[7] For example, in classical logic negation and disjunction are sufficient to generate all the other connectives. In particular:

$$A \wedge B =_{def} \neg(\neg A \vee \neg B)$$

where we read $=_{def}$ as "is by definition equal to" or "is defined as". How do we determine that this definition is adequate? Again, truth tables come in handy here.

A	B	¬A	¬B	¬A ∨ ¬B	¬(¬A ∨ ¬B)	A ∧ B
T	T	F	F	F	T	T
T	F	F	T	T	F	F
F	T	T	F	T	F	F
F	F	T	T	T	F	F

---

7. Actually, in classical logic we find something even stronger: any possible connective (having a truth-functional semantics) can be defined in terms of negation and disjunction.

5 Deductive arguments

What did we show with this table? Before answering, let us see how the table should be read. The last column and the first two ones together do nothing but present the truth table of the conjunction. Columns from the third to the sixth help to progressively construct the truth table of $\neg(\neg A \vee \neg B)$ starting from the truth-values of the sentences that constitute it – from most complex to simplest, they are $\neg A \vee \neg B, \neg A, \neg B, A, B$.

The crucial point here is indeed the sixth column: if you compare it with the last one, you see that they are the same, or in other words: $\neg(\neg A \vee \neg B)$ is true in all and only scenarios where $A \wedge B$ is true. Therefore, $\neg(\neg A \vee \neg B)$ is false in all and only the scenarios where $A \wedge B$ is false. Therefore, the two formulas $\neg(\neg A \vee \neg B)$ and $A \wedge B$ are logically equivalent – that is, the one is true in all and only the logically possible scenarios in which the second is, and false in all and only the logically possible scenarios where the second is. This shows that the definition we have proposed is *adequate*.

From here we can then go on to define other connectives. For example, xor can be defined like this:

$$A \text{ xor } B =_{def} (A \vee B) \wedge \neg(A \wedge B)$$

Again, we use truth tables to check this:

A	B	A ∨ B	A ∧ B	¬(A ∧ B)	(A ∨ B) ∧ ¬(A ∧ B)	A xor B
T	T	T	T	F	F	F
T	F	T	F	T	T	T
F	T	T	F	T	T	T
F	F	F	F	T	F	F

The last column, together with the first two, gives us xor's truth table. For the rest, what have we done? We built the truth table of $(A \vee B) \wedge \neg(A \wedge B)$ – second to last column – on the ground of the previous columns. The second last column and the last one are exactly the same. Therefore, $(A \vee B) \wedge \neg(A \wedge B)$ and A xor B are logically equivalent. This shows the adequacy of the above definition.

What the definition tells us, out of symbols, is exactly what the xor truth table tells us, that is: A xor B is true if and only if at least one of the disjoint is true, *and it is not the case that both disjuncts are true*. The difference is that the definition that we have just discussed does this in a *syntactic way*, i.e. by defining xor on the ground of other elements of the language (the connectives $\neg, \vee, \wedge$), while the truth table does it in a *semantical way*, i.e. by exploiting *extra-linguistic* entities – namely: the truth-values.

Notice that the definability of $A \wedge B$ in terms of negation ($\neg$) and disjunction ($\vee$) secures that the same xor can be defined starting from those two connectives. In particular, we have:

$$A \text{ xor } B =_{def} \neg(\neg(A \vee B) \vee \neg(\neg A \vee \neg B))$$

which tells us that A xor B is true if and only if it is not the case that: A and B are both false (that is, that their inclusive disjunction is false) or that they are both true (that is that the disjunction of their negations is false).

We will then see in Chapter 6 that the material conditional → is definable in terms of negation (¬) and disjunction (∨) – more precisely: $A \to B =_{def} \neg A \vee B$. We will also see that the biconditional ↔ is definable in terms of conditional and conjunction (∧) – more precisely: A → B ∧ B → A. Consequently, even the biconditional is, in the end, definable in terms of negation and disjunction. We will also notice that → is definable in terms of negation (¬) and conjunction (∧) – more precisely: $A \to B =_{def} \neg(A \wedge \neg B)$. Since the conjunction is itself definable in terms of negation and disjunction, this last definition becomes a more indirect way of defining the conditional in terms of those two connectives.

Notice that it is possible to start from a different ground, assuming as *primitives* the connectives of negation and conjunction. On this ground, we can then define the disjunction in a rather elementary way:

$$A \vee B =_{def} \neg(\neg A \wedge \neg B)$$

The following truth table shows us that the definition is adequate:

A	B	¬A	¬B	¬A ∧ ¬B	¬(¬A ∧ ¬B)	A ∨ B
T	T	F	F	F	T	T
T	F	F	T	F	T	T
F	T	T	F	F	T	T
F	F	T	T	T	F	F

Once we have defined ∨ in terms of ¬ and ∧, we can go on and define → via one of the two definitions we have presented already, and then define ↔ as above.[8]

We close this overview by noticing that the connectives that we have presented are the operators called "Boolean operators", or simply "Booleans", in computer science. This name is related to the British mathematician and logician George Boole (1815-1864), who first studied them. It is not by chance that these connectives play a role in computer science. Just to get an idea, the modern processors that

---

8. Notice that the logical equivalences that we have used to define conjunction and disjunction resemble the "De Morgan laws" from set theory: A ∩ B = – (–A ∪ –B) and A ∪ B = –(–A ∩ –B). This is not by chance: we can think of a sentence like *p* as denoting all the situations in which *p* is true. Thus, *p* ∧ *q* denotes the intersection of the set of situations in which *p* is true with the set of those situations in which *q* is true, and this intersection is in turn the set in which both sentences are true. Also, *p* ∨ *q* would then denote the union of the set of situations in which *p* is true with the set of those situations in which *q* is true, and this union is in turn the set in which at least one of the two sentences is true. Finally, –*p* would denote the set of those situations in which *p* is false, which is the complement of the set of situations in which *p* is true. Indeed, it is also not by chance that symbols ∨ and ∧ resemble the set-theoretical symbols ∪ and ∩ (which denote union and intersection, respectively).

make our computers and smartphones work are nothing more than a very complex system of *logic gates*. A logic gate is a digital circuit which, given the presence or absence of electric current in its inputs, allows electric current to pass through or not. For example, the AND logic gate has two inputs and one output. It passes electrical current out if it receives electrical current from *both* of its inputs. The similarity with ∧ is striking. And it is not the only one. We have the logical gate NOT, corresponding to ¬, OR, corresponding to ∨, XOR, corresponding to exclusive disjunction, NAND (i.e. NOT AND), corresponding to ¬ (A ∧ B), etc.

Truth conditions and the validity of arguments

The notion of *validity* is a semantical notion, since it crucially relies on the notion of *truth*. Recall that an argument is *deductively valid* if and only if its conclusion is true in every logically possible scenario in which its premises are true. The semantics of propositional logic specifies the truth conditions of complex sentences that contain sentential operators. The truth tables give us a very simple visual representation of these conditions.

These two points helps us check the *validity* of an argument in a systematic and rigorous way. In order to understand why, consider that the rows of a truth table represent in a certain sense *all the logically possible scenarios* in which the constituent sentences have a given truth-value, and by scrolling through the rows, we can check which truth-value the complex sentence that we wish to evaluate must have, in those scenarios. For example, in the disjunction truth table, the last row (down) ideally represents all possible scenarios where A and B are false, and tells us that, in all of these scenarios, A ∨ B is itself false. Due to this, truth tables – and more generally, propositional semantics – give us everything we need to check the validity of an argument whose sentences are simple or contain only truth-functional sentential operators. Indeed, if we are dealing only with such sentences, the *connectives* are the key to reasoning – or else: connectives are what determines the rational connections between sentences. And semantics tells us under what conditions the connectives "generate" true sentences. Let us take an example. Take the following argument:

(P1)　　A ∨ B
(P2)　　¬B
―――――――――
(C)　　　A

This argument is valid, and the truth tables help us see why. Take the table:

A	B	A ∨ B	¬B
T	T	T	F
T	F	T	T
F	T	T	F
F	F	F	T

The first three columns together form the truth table of disjunction. The fourth column gives us the truth conditions of ¬B in relation to the four possible scenarios defined by the combination of the truth-values of A and B. The rows of the table represent precisely these possible scenarios. To see if the above argument is valid, what should we do, after building the table above? We should go and see the rows where *both* A ∨ B and ¬B are true. If A is true in all of these scenarios, then the argument is valid. Otherwise, it is not. There is only one group of scenarios in which both A ∨ B and ¬B are true, and this group of scenarios is represented by the second row (with the exclusion of the headline, which lists the sentences). In that row, B is true. The argument is therefore *valid*.

The following table gives us an alternative visual representation of the same control procedure. It starts directly from the four possible combinations of truth-values of the two sentences that are indirectly involved in the two premises (i.e., A and B).

Combination	P1: A ∨ B	P2: ¬B	C: A
TT	T	F	T
TF	T	T	T
FT	T	F	F
FF	F	T	F

Notice that since we have *two* sentences and *two* possible truth-values, we have $2^2 = 4$ possible combinations (ideally, four rows, as in the truth tables that we have seen thus far). If we had *three* sentences and *two* truth-values, we would have had $2^3 = 8$ combinations. If we had *two* sentences and *three* truth-values, we would have had $3^2 = 9$ possible combinations. In general, in a truth table we have m^n possible combinations of truth-values, where m is the number of truth-values that can be assigned, and n is the number of the simplest sentences occurring in the sentences under consideration.

The inference scheme that we have just discussed is a scheme that we use very often in everyday life. If we know that Anna is home or at the gym and if we find out that she is not home, then we conclude that she is at the gym. If I can choose between two options and discard one because it leads to a false consequence, I will choose the other one.

We can give a general recipe to apply the control method that we have just illustrated with the simple example above. It consists of the following three steps:

1. Identify all the combinations of truth-values of the sentences that appear in the premises and in the conclusion;
2. Build a table in which each row represents a combination and reports the truth-values of the premises and conclusion for those combinations;
3. Verify that there is no row in which for all the premises we have value T and for the conclusion value F. If there is no such row, the argument is valid. However, if such a row exists, the argument is invalid.

5 Deductive arguments

We also exemplify the semantic proof of the *invalidity* of an argument. Take:

(P1)   A ∨ B
(P2)   A
─────────────
(C)   B

This is *not* a valid argument, and we just need to consider the truth table of disjunction in order to see this:

A	B	A ∨ B
T	T	T
T	F	T
F	T	T
F	F	F

The second row represents the family of possible scenarios in which A ∨ B is true, A is true, and yet B is not. This gives us a possible scenario where P1 and P2 are true, but C is not: the inference scheme in question is invalid.

In Chapter 6 we will apply the same procedures and methods to prove the validity or invalidity of inference schemes involving the material conditional →.

Before closing this paragraph, let us show the *validity* of A ∨ ¬A. Take the following table:

A	¬A	A ∨ ¬A
T	F	T
F	T	T

Naturally, A determines two (families of) possible scenarios: those in which A is true (and ¬A is false), and those in which A is false (and ¬A is true). Given the truth tables of ¬ and ∨, we have that A ∨ ¬A is true in both families of scenarios. But since either A is true, or A is false, A ∨ ¬A is true in every logically possible scenario. When a sentence A is true in any logically possible scenario, it is said to be a *valid* sentence. This extension of the term "valid" from inferences to sentences is due to the fact that, if A is true in every possible scenario, then any inference that concludes A is valid, no matter what premises it starts from.[9] Another way to say the same thing is that A is a *logical truth* (or, in the case of propositional logic, a *tautology*).

---

9. Indeed, if A is true in every possible scenario, then it will be true in every possible scenario in which $A_1, ..., A_n$ are true – with n finite and $1 \leq n$, and for arbitrarily chosen $A_1, ..., A_n$. This is due to the fact that, if A is true in every logically possible scenario, then any inference that concludes A from any premise B is valid.

5.2.3 Proving by rules. Natural deduction

Semantics specifies the truth conditions of sentences and allows us to say whether an argument is valid or not. This is a very intuitive way of understanding how an argument works, and it is also the most natural way for us, since the distinction between a good argument and a bad argument (which is central to this volume) depends partly, albeit crucially, on whether all the premises are *true* or not. However, there is also a way of understanding the functioning of an argument that is independent of the notion of truth, and is, as logicians say, purely *syntactic* – that is, it focuses on the kind of logical symbols that appear in the sentence, and not on the truth of the sentence. The central notion in this approach is that of a *proof.* A proof can be seen as a process that takes a series of sentences (the premises) as input, and outputs a sentence (the conclusion). In a proof, the "input" sentences are *assumed* or *derived* from sentences already given as input through *inference rules*. The rules of inference also help us go from the inputs (the premises) to the output (the conclusion), and in this volume we focus mainly on this role of the rules. Inference rules can be thought of in terms of validity (which would bring us back to the semantic angle) or simply as (non-arbitrary) instructions to *manipulate* sentences that contain certain logical concepts, such as connectives (in the case of rules for propositional logic we will see shortly what this means in concrete terms). In other words, in a proof everything that appears as a premise is either an assumption, or it has been previously proved based on other sentences.[10]

Although in this volume we rely on a semantic understanding of deductive reasoning procedures (and of reasoning in general), it is also worth taking a look at the syntactic approach, because it gives us the chance to see reasoning in terms of *rules* rather than in terms of truth. In what follows, we will briefly discuss the

---

10. Ideally, what is simply assumed should always be "discharged", through a process notifying that the assumption did not necessarily came with a commitment to the truth of the assumed sentence, or from the fact that a proof of that sentence has been given already or will ever be given. We will consider an example of discharging in Chapter 6, when we will discuss deductive hypothetical reasoning. Indeed, discharging is essential in what are termed "hypothetical derivations" in which that kind of reasoning consists. Discharging what has been assumed without first being proved is crucial to ensure that what appears in a proof is always well-grounded, in conformity with the need for rigor that comes with mathematical reasoning – it is indeed in connection with this species of reasoning activity that the notion of proof has been first defined in a rigorous and systematic way. Of course, we do not aim at the same degree of rigor as mathematical proofs in our everyday reasoning, but in a similar way we can say that, if we do not offer any reason for the truth of our premises, the best we can do is, actually, prove the logical connections between such premises and the conclusion (which is what we ultimately do when we discharge an assumption at the end of a proof). In the best situations, we can take the premises for granted if they have already been accepted by the other parties that are involved in the argument exchange in which we are involved. From this point of view, similar dynamics apply to mathematical reasoning and to everyday reasoning, although the latter proceeds inevitably on looser criteria.

idea of reasoning through rules concerning connectives, and we will present, albeit briefly, what is termed "natural deduction", which gives us an example of reasoning through rules and exemplifies in a very simple way the approach to reasoning that does not need to call into question the concept of truth.

Proving by rules

The scheme of inference

$$\frac{A \lor B \quad \neg B}{A}$$

gives us a *rule* of reasoning that sounds like this: whenever at a given point of your argument you have a disjunction (A ∨ B) and the negation of one of its disjuncts (¬B), conclude the other disjunct (A). Notice that this explication of the rule does not involve the concept of truth, but simply what we can do if the sentences A ∨ B and ¬B appear as "inputs" of our reasoning.

Rules such as the one we have just exemplified are called "rules of inference". These rules are not arbitrary, and there are two ways to understand this point. One way is to say that the "right" rules of inference are those that are deductively valid. This brings us back to semantics, and in particular gives us a semantic criterion for distinguishing inference rules worth their salt from pseudo-rules, that is, those deductive reasoning schemes that are fallacious. Another is to say that an inference rule is worth its salt if it can be derived from a basis of "primitive" or elementary rules that are for some reason secured – most importantly because they detail the basic behavior of the individual connectives we have introduced.

This implies the construction of a "structure" of inference rules in which there are a few rules that "generate" all the others in some sense. There are many different ways of implementing this idea formally, and thus of defining a propositional calculus based on rules. We will briefly illustrate one of them, known as *natural deduction*.

Before doing so, notice that the idea of starting with a sufficient number of primitive rules alone to generate *all* the rules of reasoning is one of two options in the syntactic approach to propositional logic (and logic in general). The other option usually starts out from a large number of *principles* (which are not rules, but sentences that are taken as axioms) and few *inference rules* (which typically include *modus ponens*, which we will look at in Chapter 6; sometimes this is the only rule); from these, it is possible to generate all the theorems of the logic. However, proofs are usually longer under the second approach. Both options fall into the branch of mathematical logic that is called *proof theory*, whose starting point is the possibility of conceiving of proofs as complex syntactic objects.

Natural deduction

In principle, we could play with an infinite number of inference rules, but we tend to use a very small pack of rules in our reasoning. The system called "natural deduction calculus", however, frames inference rules in a very simple way, as it individuates *two* inference rules for each logical operator (or, more in general, two *kinds* of rules). These are termed *introduction rules*, telling us under which conditions we can introduce a sentence having a given main connective in a proof, and *elimination rules*, telling us under which conditions we can go, in a proof, from a sentence having a given main connective to a sentence that does not contain that connective.

Natural deduction for propositional logic provides a very simple approach to rules, and we will briefly go through it. Before doing so, it is worth making a short comment: we owe natural deduction calculus to the German logician Gerhard Gentzen, who presented a natural deduction calculus for classical logic. Gentzen called it "natural deduction" in the belief that its rules were the rules that mathematicians actually used ("naturally") in their reasoning. It was later questioned whether mathematicians actually reasoned in the way Gentzen described – and apparently for good reason – but the name of the calculus has been left as it was.

Let us take conjunction (∧). This is the *rule of conjunction introduction*:

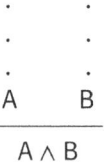

The rule tells us that if we have got to a sentence A along a proof (or we assumed A along the way), and we have got to a sentence B along a proof (or we assumed B along the way), then we can (legitimately) conclude A ∧ B. By doing this, the rule specifies the exact conditions under which we can *introduce* a conjunction along a proof, or, more generally, under what conditions we are entitled to conclude a conjunction in an argument.

Note that A and B, above the line, are on two different lines because we do not need them to be in the same proof or argument – although this is possible. Summing this up: if A has been proved previously, and B has been proved previously (even in different proofs), then it is also implicitly proved that A ∧ B. More generally, if A follows from one of our arguments, and B follows from one of our arguments, we are entitled to conclude A ∧ B. The semantics introduced in § 5.2.1 helps us check the *validity* of the rule. Indeed, if we take the truth table of the conjunction, we will see that the row in which both A and B have value T is such that A ∧ B also has value T.

The following rule is known as the *rule of conjunction elimination*:

$$\frac{A \wedge B}{A}$$

The rule tells us that if I have a conjunction along my reasoning, I can (legitimately) conclude one of the two conjuncts (arbitrarily chosen). The semantics introduced in § 5.2.1 helps us check the *validity* of this rule. Indeed, A ∧ B is true in only one case, the one in which both A and B are true. Note that the presence of A rather than B above is arbitrary. The rule would remain valid even if we illustrated it with the other conjunct.[11] As we can see, the rule tells us in what circumstances, along a proof or an argument, we can "eliminate" a conjunction – that is, use the conjunction to move to a simpler sentence (any of the conjuncts).

The *rule of disjunction introduction* is:

$$\frac{A}{A \vee B}$$

That is: where you have a sentence along a proof, you can (legitimately) introduce any disjunction such that the sentence in question is one of the disjuncts. That this is a valid rule is clear from the truth table of the disjunction: A ∨ B is true in any row in which at least one of A and B is true.

It has been questioned whether the disjunction of our ordinary language actually works like this. After all, the rule tells us that we can introduce *any* disjunction in which a sentence occurs that we have previously assumed or proved. Now suppose I correctly conclude from the basic laws of arithmetic that 2 + 2 = 4, and then go on to say, "So, 2 + 2 = 4 or Joe Biden is the President of the USA". This would sound strange, and for two reasons. The first reason is that the procedure leading to a proof that 2 + 2 = 4 does not contribute at all to a possible procedure to "prove" that Biden is the President of the USA. The second reason is that "2 + 2 = 4" is completely irrelevant to "Joe Biden is the President of the USA".[12]

In any case, as long as we accept the truth table for the disjunction introduced in § 5.2.1, the rule of disjunction introduction is *valid*. The rule of *disjunction elimination* is more complex:

---

11. It is possible, and in a sense more correct, to define *two* rules of conjunction elimination: one allowing us to infer A from A ∧ B, and one allowing us to infer B from A ∧ B. Since conjunction is commutative, however, we can also specify just one rule, as we are doing here. A similar remark holds for the rule of disjunction introduction.
12. We will touch upon the issue of relevance in Chapter 6, when we will discuss the material conditional.

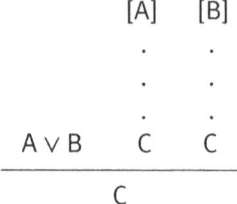

The rule tells us that, if we have proved or assumed A ∨ B, and C can be derived *both under the assumption* of A *and under the assumption of* B, then we can conclude C. "Derivable under the assumption of A and B" means that A and B are assumed exclusively to see what follows (the use of square brackets indicates this).[13] The sense of the rule is this: if a sentence follows both from A and from B, then the disjunction A ∨ B suffices to conclude that sentence. In terms of truth-values: if C follows as much from A as from B, then C is true in every possible scenario where A is true, and in every possible scenario where B is true. These are, taken together, the possible scenarios where A ∨ B is true. The rule is therefore *valid*.

Let us look at *the rule of negation introduction*:

This is equivalent to saying that if we derive a contradiction from a sentence, we can (legitimately) conclude the negation of the sentence itself. In essence, this is the *reductio ad absurdum* that we discussed in Chapter 4. In terms of truth-values, we can analyze the rule like this: *if* the truth of A guarantees the truth of a contradiction, then A is false. Indeed, "the truth of A guarantees the truth of a contradiction" equates with saying that a contradiction is true in every logically possible scenario in which A is true. But there is no possible scenario in which a contradiction is true, and therefore there is also no possible scenario in which A is true. Hence, given the truth table of negation, ¬A is true in every logically possible scenario. The one indicated below is instead termed the *rule of negation elimination*:

$$\frac{\neg\neg A}{A}$$

---

13. In particular, the use of square brackets signals that discharging assumptions is essential in order to fully understand the rule correctly.

According to this rule, if we prove the negation of a negation, we automatically prove the negated sentence. The truth table for negation makes it clear that ¬¬A is true if and only if A is true:

A	¬A	¬¬A
T	F	T
F	T	F

We also have introduction and elimination rules for the conditional → , which we will look at in Chapter 6. The rules for the biconditional follow from them, and from the rules for conjunction, in an elementary way. By adding the rule:

$$\frac{}{A \vee \neg A}$$

to the pack, we get the calculus of natural deduction for *classical* propositional logic. The rule in question tells us that we can always introduce, at any point in a proof, a logical truth such as A ∨ ¬A. The reason why natural deduction is interesting is that the simple and compact set of rules that we have introduced is able to generate *all and only the reasoning schemes* that are *valid in classical propositional logic*. In other words, it constitutes an economic and compact basis from which any form of deductively valid inference in this logic can be constructed.

5.2.4 Assertion and rules of inference

What we have seen so far about natural deduction gives us a rather abstract perspective on deductive reasoning – or rather, on that part of deductive reasoning in which connectives are the only logical terms. Let us try to reach a less abstract (and hopefully intuitive) perspective on the above rules of inference. We can do this by considering some of our linguistic practices concerning *assertion*. This will help us approach the rules via a very familiar phenomenon that we experience every day: that of asserting sentences, that is, less technically, making statements. Let us illustrate this with an example.

If someone asserts "It is sunny outside and it is twenty-five degrees", we are entitled to expect that, at least in principle, that person is also willing to assert, or defend, *both* "It is sunny outside" *and* "There, it's twenty-five degrees". Imagine this exchange of words between Benedict and Alice, with Anna appearing just as a listener:

> *Alice*: It's sunny outside and it's twenty-five degrees.
> *Benedict*: Anna, did you hear? It's sunny outside!
> *Alice*: No, no, absolutely not! It's not sunny outside.

Alice's verbal reaction seems strange to us, because she has first *asserted* a conjunction A ∧ B and then she has *denied* one of its conjuncts – more precisely: she *has asserted* the *negation* of one of them. The reason why this sounds strange is simple: the *assertion of a conjunction ideally commits the speaker to separately asserting each of the conjuncts*. What does "ideally commit" mean? It means in the first place that whoever listens to that assertion is entitled to expect that whoever asserts it also asserts, if requested by the others, any of the conjuncts, or that the speaker supports the assertion of one of the conjuncts, if such assertion is made by others. For example, if Benedict had said: "It's sunny outside. Do you confirm this, Alice?" it would have been legitimate for him to expect Alice to say, "Yes, I do" (while maybe also giving some kind of reason for her statement). By the same token, in the example above Benedict is entitled to be surprised: Alice has asserted a conjunction but she is not defending one of the conjuncts – indeed, she has even asserted its negation. All this means that we cannot (i) assert a conjunction, (ii) deny one of its conjuncts, or assert its negation, or refuse to support it or give it our approval, and (iii) be *rational*. In other words: if a *rational* speaker asserts a conjunction, then they ideally commit themselves to asserting (defending, accepting) each of the conjuncts. If a listener assumes that the speaker is rational, then they are entitled to expect the speaker to fulfil the commitment.

Clearly, this provides an intuitive understanding of the rule of conjunction elimination: if in your argument you have asserted (as an assumption or consequence of an assumption) A ∧ B, then ideally you commit yourself to asserting A as well, and also B (and at the same time, you make the assertion of each conjunct legitimate). If the listener accepts your premise A ∧ B, then they also commit themselves to accepting A and accepting B.

Something similar holds for conjunction introduction: if I assert A, and I assert B (in two different acts of assertion), I ideally commit myself to asserting A ∧ B. It would be irrational to assert (or accept) two sentences and refuse to accept their conjunction.

Let us briefly consider how the commitments generated by assertions help us understand the rules for disjunction and negation. If I have asserted A ∨ B, and then in one argument I validly conclude C from A, and in another argument I validly conclude C from B, then I ideally commit myself to asserting C, and at the same time I legitimate myself in asserting C. It would be irrational to assert a disjunction, to show that from any of the disjuncts C follows, and yet reject C. This allows us to have an intuitive understanding of *disjunction elimination*. As for *disjunction introduction*, it is at odds with our intuitions, as we have seen above. Let us try to understand it like this. The "classic" view we are following suggests that, if I have asserted A, it makes no sense that I refuse to accept A ∨ B, no matter what B is, because refusing to accept that disjunction is tantamount to refusing to accept *both* disjuncts. But I have just asserted one of them. This argument assumes that "refusing to accept a disjunction" is equivalent to "refusing to accept both disjuncts". If the disjunction works as in the truth table for ∨ that we have seen in § 5.2.1, this equivalence actually holds – and since the truth of A suffices to secure the truth of A ∨ B, the acceptance of A rationally justifies the acceptance of A ∨ B.

If I assert A and I derive a contradiction from this, it is rational that I assert the *negation* of A. Otherwise, I should be ready to assert also what I have derived from A, that is a contradiction. But then I would not be rational. If I assert that it is not the case that A is not the case, it is rational for me to assert A.

These comments enable us to read the rules of natural deduction in a way that is more suited to our daily argumentation activity – in fact, reading them in terms of "proof" makes more sense in mathematical reasoning. For instance, if discussing mathematical reasoning, we would read conjunction introduction in the "official" way in which logicians read it, namely: "If you have proved (or assumed) A, and you have proved (or assumed) B, then you have proved A ∧ B". What we have said about assertion suggests that we can read the rules in a more general way, namely: "If you have asserted (or assumed, or concluded) A, and you have asserted (or assumed, or concluded) B, then you are entitled to conclude A ∧ B" (and at the same time, you are also ideally committing yourself to concluding such a conjunction).

5.3 Reasoning with predicates and quantifiers

Let us go back to the three deductively valid arguments that we exemplified at the beginning of the chapter.

(P1) All sculptures imitate natural objects.
(P2) *Bird in space* does not imitate a natural object.
___
(C) *Bird in space* is not a sculpture.

(P1) All sculptures imitate natural objects.
(P2) *Unique forms of continuity in space* does not imitate a natural object.
___
(C) *Unique forms of continuity in space* is not a sculpture.

(P1) All horses have four legs.
(P2) Furia does not have four legs.
___
(C) Furia is not a horse.

Let us try to understand what kind of inference scheme we are facing here – or equivalently, to analyze the logical form of these three arguments.

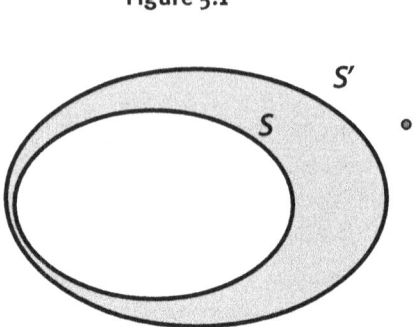

Figure 5.1

Premise P1 tells us that a set S of objects is included in another set of objects S'. Premise P2 tells us that an object i *does not belong to* set S'. Conclusion C says that i *does not belong to* set S (Figure 5.1). This conceptual structure is common to all three arguments. Moreover, the arguments are valid *precisely* due to their conceptual structure. The relations between the sets S and S' and the object i make the conclusion true, given the truth of the premises. In other words, it would be inconsistent to say that a set S is included in a set S', that i does not belong to S' and that, nevertheless, i belongs to S.

In short, the validity of those three arguments does not depend on which predicates appear in them (and which specific the sets of objects they *refer to*), but on the type of relationship between the sets of objects that are expressed by the premises and the conclusion. Therefore, not only these three arguments, but *all arguments presenting this logical form* are valid: if an argument presents this logical form, the truth of the conclusion is guaranteed, given the truth of the premises.

Recall, before we go on, that a deductively valid argument does not necessarily have a true conclusion, and that if (some of) its premises are false, the conclusion may also be false. This in no way jeopardizes its validity. Let us take for instance:

(P1)   All horses have four legs.
(P2)   Furia does not have four legs.

(C)   Furia is not a horse.

Not all horses have four legs, because there will be at least one that has lost a leg in a bad accident. Hence, P1 is false, and knowing this keeps us from concluding C, even if we take P2 as true. However, the argument is deductively valid, because if P1 and P2 were both true, then C should also be. We can schematize the three arguments discussed above as follows:

(P1)   All *P*s are *Q*.
(P2)   *c* is not *Q*.

(C)   *c* is not P.

When nouns and predicates are relevant to our reasoning (contrary to the propositional case), it is even more clear that the schematic letters help highlight the logical structure of the argument. In fact, while we have replaced the particular nouns and predicates used in the arguments with schematic letters, we have preserved the words that tell us what the logical relationships between the sets and the individuals denoted by these nouns and predicates are (in particular, we have kept the word "All"). As we have seen, it is these relationships that are crucial in the above three arguments.

Since the validity of the argument is guaranteed by its logical form, we can use P, Q and c respectively in order to stand for any common nouns, any predicate, any proper name and we will still obtain a valid argument, since the specific identities of the sets to which the predicates somehow refer, and the identities of the individuals to whom the nouns refer, are *irrelevant* to the validity of the argument.

5.3.1 Sentences containing predicates and quantifiers, and their relationships

As we have seen in the previous section, usually our sentences are composed of a subject and a predicate. An important class of sentences in subject-predicate form relates the set of objects denoted by the subject with the set of objects denoted by the predicate. For example, if I say that *all horses have four legs,* I am relating the set of horses (or as we will say, the *set of those individuals who are horses*) with the set of individuals who have four legs. Sentences that relate two sets of individuals are called *categorical*, precisely because they concern categories of individuals and their relationships – at least in a loose and intuitive sense of "category". For example, the following are categorical sentences:

(1) All sculptures imitate natural objects
(2) All horses have four legs
(3) All elephants are equiangular
(4) No senators have been bribed
(5) Some TV stars care about their privacy
(6) Not all movies have a happy ending
(7) Some students will not pass the exam

The first three sentences say that one set is included in another, and their schematic representation is: All Ps are Q. The fourth sentence says that the intersection between the set of senators and that of corrupt people is empty, and its schematic representation is: No P is Q – or, equivalently: All Ps are not Q. The fifth sentence says that the intersection between the set of television stars and that of the people who care about their privacy is not empty, and its schematic representation is: There exists some P which is Q – or, equivalently: Not all Ps are non-Q. The sixth sentence says that the intersection of the set of movies and the complement of the set of happy ending stories is not empty, and the seventh sentence says the same with respect to the students and those who will pass the exam. Their form is: There exists some P that is not Q.

From a set-theoretical point of view, the relationships that these sentences establish between the two classes are the following, where S is the set of individuals to which the predicate P refers, and S' that of the individuals referred to by the predicate Q:[14]

$$S \subseteq S'$$
$$S \cap S' = \emptyset$$
$$S \cap S' \neq \emptyset$$
$$S - S' \neq \emptyset$$

The words *all, none, exists/some* are called *quantifiers* because they indicate how much of class S is included in or excluded from class S'. Note that sentences like these reduce all predicates to the form "copula + noun", even when this copula is not present in the original sentence. For example, the predicate "having four legs" is rendered as "being an individual that has four legs". This paraphrase of all categorical sentences with the general form:

Quantifier *P* is (not) *Q*

dates back at least to Aristotle, and it has generated a huge number of discussions among logicians and philosophers of language. Here, however, we will disregard these problems and assume that the above is a correct periphrasis, which does not distort the meaning of the sentence. We will say something more about the periphrasis in § 5.3.3.

During the Middle Ages, the four forms of categorical sentences were traditionally indicated using the four letters **A, E, I, O**, which we write here in bold so as not to confuse them with the schematic letters for arbitrary formulas[15]. The quantifier "All" (or "For every") is called the *universal quantifier*, and the sentences of kind **A** and **E** ("All *P*s are *Q*", "No *P* is *Q*") are called *universal*, while the quantifier "Some" (Or "There is at least one") is called the *existential quantifier* and the sentences of kind **I** and **O** ("Some *P*s are *Q*", "Some *P*s are not *Q*") are called *existential* or *particular*. The relationships between these four kinds of sentences are illustrated by what is traditionally called the square of oppositions (Figure 5.2).

---

14. Notice that what we are calling, here, "the set of the individuals to which predicate *P* refers" is the set of those objects to which we refer by what we call "subject" (of a quantified sentence) in our natural language. What we are calling "the set of the individuals to which predicate *Q* refers" is the set of objects to which we refer by what we call "predicate" (of a quantified sentence) in our natural language. For instance, in "Every horse is black", the subject is "horse", the predicate is "is black". We will see that formal logic interprets sentences of this kind as sentences that contain no subjects, but just predicates ("to be a horse", "to be black") and individual variables.

15. It is thought that the labels derive from the Latin words *AffIrmo* e *nEgO* because "All *P*s are *Q*" and "Some *P*s are *Q*" are sentences that assert something while "No *P* is *Q*" and "Some *P*s are not *Q*" are sentences that deny something.

Figure 5.2 The square of oppositions

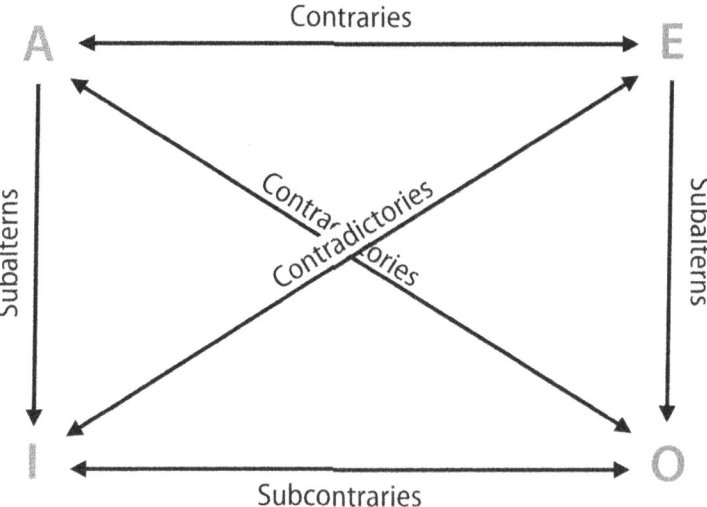

Sentences **A** and **E** are contraries. Two sentences are said to be contraries when they can both be false but they cannot be both true. Thus "All students passed the exam" and "No student passed the exam" can be both false (if some students passed the exam and some students did not pass) but they can never both be true: if one is true the other is necessarily false.

A and **I** on the one hand and **E** and **O** on the other are subalterns. Two sentences are subalterns when they imply each other. In particular, **A** implies **I** and **E** implies **O**: if it is true that all the students have passed the exam, it will of course also be true that some students have passed it. If it is true that no student has passed the exam, it is of course also true that some students have not passed it. If this may seem strange, think of set-theoretical relations: if the set S is included in the set S', this implies that the intersection between S and S' it is not empty; if the intersection between S and S' is empty, then the set $S - S'$ will not be empty (if S is not).

Sentences of kinds **I** and **O** are subcontraries. Two sentences are subcontraries when they can both be true but they cannot both be false. If some students passed the exam and some students did not pass, then an **I**-sentence and a **O**-sentence are both true, but it cannot be false that some students passed the exam and false that some students did not. In other words, it is not possible that there are neither elements of S which are S' nor elements of S which are not S'. The subcontrariness of **I** and **O** can be verified by means of the relations we have already described. For example, we have said that **A** and **E** cannot both be true, and this excludes that their contradictories are both false because when **A** and **E** are true, **I** and **O** are false. To understand the relationships between sentences, a set-theoretical representation may prove helpful. A sentence **A** states that the set of S is included in that of S' (Figure 5.3).

Figure 5.3   **Figure 5.4**

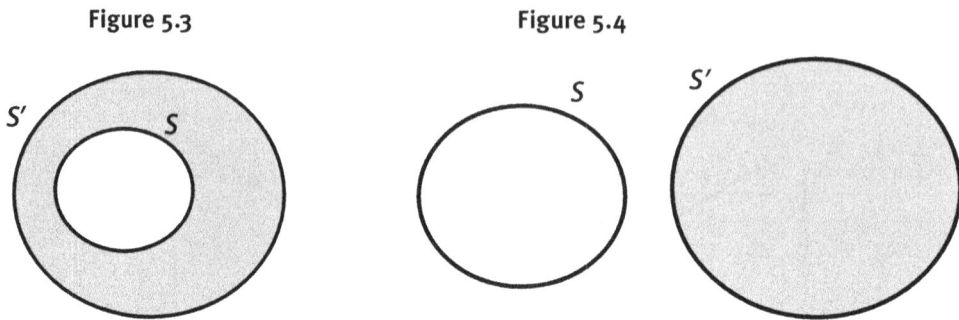

If this is true, then it will be false that the intersection between the sets S and S′ is empty and therefore the relevant **E**-sentence will be false. It will also be false that there are elements of S that are not elements of S′ and therefore also the relevant **O**-sentence will be false. Instead, it will be true that the intersection between S and S′ is not empty and therefore the relevant **I**-sentence will be true.

An **E**-sentence states instead that the intersection between the sets S and S′ is empty (Figure 5.4). If this is true, then it will be false that the set S is a subset of the set S′ and therefore the relevant **A**-sentence will be false. It will also be false that there are Ss that are S′s – that is, that the intersection of the two sets is non-empty – and therefore the relevant **I**-sentence will be false. Instead, it will be true that there are elements of S that are not elements of S′ and therefore the relevant **O**-sentence will be true.

A sentence of kind **I** tells us that the intersection between S and S′ is not empty (Figure 5.5). The asterisk in the figure indicates that the set in question is non-empty. If this is true, then it will be false that the intersection between the two sets is empty and therefore the relevant **E**-sentence is false. However, in this case we can say nothing of **A** and **O**. Indeed, a sentence of kind **I** just says that there are elements in S ∩ S′ but it does not say whether all the elements of S are in S′ or not or if all the elements of S′ are in S or not. So we do not know if **A** and **O** are true or false.

Finally, **O** tells us that there are elements of S that are not in S′ (Figure 5.6). If this is true, then the set S is not totally included in S′ and therefore the rele-

Figure 5.5   **Figure 5.6**

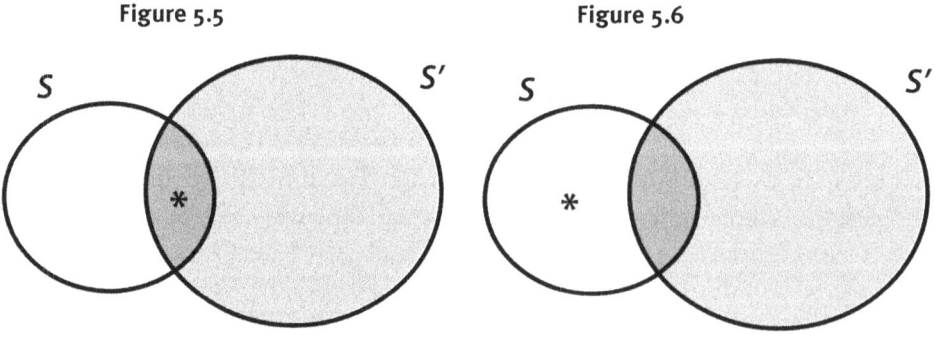

vant **A**-sentence will be false. But we cannot say anything about the relevant **I**- and **E**-sentences. Indeed, a sentence of kind **O** just says that there are elements in $S - S'$ (if S is non-empty) but we do not know if all the elements of S are outside S' or not. So we do not know if **I** and **E** are true or false.

The relationships that we have illustrated here between the categorical sentences have been challenged, since today it is believed that the sentences **A** and **E** *do not imply* that the sets S and S' are *non-empty* and therefore such sentences can be true even when such sets do not contain any elements. For example, suppose you read the following warning in public transport: "Do not smoke. All transgressors will be punished according to the law". The sentence that all offenders will be punished according to the law is true even if there is (and never will be) any offender. Hence an **A**-sentence can be true even if the set of objects denoted by its subject is empty. Conversely, the sentences of types **I** and **O** seem to have an existential implication, that their truth implies the existence of objects in S. "Some Martians are green" does not seem to have any chance of being true if there is no Martian. Therefore, if an **A**-sentence can be true even if the corresponding **I**-sentence is not (and the same applies to types **E** and **O**), we can no longer say that **A**-sentences imply **I**-sentences I and that **E**-sentences imply **O**-sentences. Here we will disregard these difficulties simply by assuming that the relevant sets for a given quantified sentence are non-empty.[16]

5.3.2 A formal language for reasoning with predicates and quantifiers

Before moving on, let us briefly consider how we can provide a rigorous artificial language that can express the sentences appearing in our reasoning with predicates and quantifiers. This will build over the language that we have provided in § 5.2 for that piece of language which is involved in propositional reasoning. Then, we will explain how to attribute a truth-value to sentences containing predicates or quantifiers, and we will briefly present the introduction and elimination rules for the quantifiers. Before doing so, we need to understand how logic "reads" sentences in subject-predicate form. Take a sentence like "Peter runs". In logic, the predicate "run" (like *any* predicate) is interpreted as a *function* that applies to a name: if P stands for "to run", and a stands for "Peter", then $P(a)$ stands for "Peter runs". This also works for relational predicates like "Lorenzo loves Giovanna", but these predicates take two (or more) names as arguments. $P(a,b)$ can therefore stand for

---

16. Notice that the idea that **A** and **E** have no existential implications gives rise to oddities. For example, "All Martians are green" is true even though there is no Martian. We can of course discuss why "All offenders will be punished" sounds true to us even if there is no offender and "All Martians are green" sounds false to us if there is no Martian, but this would take us too far from the aim of this volume.

"Lorenzo loves Giovanna". As for quantified sentences, words like "All", "Some", "There is at least one", "For every", are treated as operators, and a sentence like "Everyone has red hair or hair that isn't red" is rendered as $\forall x(P(x) \vee \neg P(x))$. A sentence such as "There is at least one city" is rendered as $\exists x P(x)$. We will see shortly how sentences such as "All men are mortal" and "Some men are tall" are rendered.

Let us now turn to our "new" language. In order to build it, we must just adjust the recursive definition of § 5.2 so that it also includes sentences of the form $P(a)$, $\forall x P(x)$, and so on. The recursive definition for formulas thus becomes:

1. A string of symbols of the form $P(a)$ is a sentence.
2. If A is a sentence, then so is ¬A.
3. If A, B are sentences, then so is A ∧ B.
4. If A, B are sentences, then so A ∨ B.
5. If A, B are sentences, then so is A → B.
6. If A, B are sentences, then so is A ↔ B.
7. If A is a sentence, $\forall x A(x)$ is a sentence.
8. If A is a sentence, $\exists x A(x)$ is a sentence.
9. Nothing else is a sentence.

Notice that sentences of the form $P(a)$ are now the simplest, and thus they play the role that atomic sentences $p, q, r, s, \ldots$ previously played. The great difference is that the sentences of the form $P(a)$ are able to reveal the basic syntactic structure of *our* simplest sentences (that is, of the sentences that are the simplest among those we use in our natural language). Steps 2) to 6) are exactly the same as in the definition from § 5.2. Step 7) tells us that if A is a sentence (be it atomic or complex), then if we replace the name – or one of the names – that appears in A with a variable, and quantify it by ∀, then we get a sentence. Step 8) says the same thing, relative to the existential quantifier. It is very important that arbitrary formulas appear in these two passages, because the non-quantified sentences on which we ideally work to obtain quantified ones can have arbitrary complexity. In other words, we do not just have universal sentences of the form $\forall x P(x)$, but also sentences of the form $\forall x(P(x) \wedge Q(x))$ or $\forall x(P(x) \rightarrow Q(x))$.

Let us now see how we can render sentences such as "All men are mortal". What does that sentence say? It says that everything that is a man is also mortal. That is, *for everything,* if *that thing is a man,* then *that thing is mortal*. In other words, sentences of the form "All Ps are Q" are read in logic as $\forall x(P(x) \rightarrow Q(x))$. The logical symbols that will determine the behavior of this kind of sentence in reasoning will therefore be those that express our "All" and "If ... then". What does a sentence like "Some cars are blue" (or "One or more cars are blue") tell us? It tells us that there is something that is a car and it is blue. In other words, sentences of the form "Some P is Q" are read in logic as $\exists x(P(x) \wedge Q(x))$.

Truth conditions of sentences containing predicates and quantifiers

It is time to move on to the truth conditions of sentences of the form $P(a)$, $\forall x A(x)$ or $\exists x A(x)$. We already have a solid intuition regarding these conditions, but let us first see how formal logic defines them, and we will then realize that logic does nothing but make our intuitions clear and rigorous. Once we have names, predicates, and quantifiers, our semantics no longer has to tell us just under what conditions a sentence is true or false. It must also tell us to which individuals (objects) the names refer, and to which sets of things the predicates refer, in some sense. To do this, we must use the concept that logicians call "interpretation".

An interpretation is a function that can take arguments of different types. A crucial point is that, for it to work correctly, we must define a precise "domain of discourse" (D) that gives us the individuals to whom or which the names refer and the sets of individuals to which the predicates refer. For example, an interpretation applies to individual constants (names), and associates with each of them the individual member of D to which the name refer (the "reference" of the name). For example, given an interpretation I and an individual constant a, we have $I(a) = d$, where $d \in D$ (d belongs to D), for some individual in D. An interpretation, however, also applies to predicates, associating to each of them a subset S of D. For example, we will have $I(P) = S$ for some set S such that $S \subseteq D$. In other words, names *denote* individuals (according to the association dictated by the interpretation), and predicates *denote* the extension of a property (that is, the set of objects that satisfies that property). For example, the predicate "to run" denotes, in this logical framework, the set of those individuals that run. An interpretation also applies to sentences and associates them with a truth-value (in our case, T or F). Here are the truth conditions of a sentence of the form $P(a)$:

$$I(P(a)) = T \text{ if and only if } I(a) \in I(P)$$

That is: $P(a)$ is true if and only if the individual to which a refers belongs to the set extending the property expressed by P. For example, if $P(a)$ is "Peter runs", then $P(a)$ is true if and only if the name "Peter" refers to an individual who is in the set of those individuals who run. Once we take it outside logical formalism, the idea is very intuitive. As for the connectives of this formal language, they are defined by the truth tables of § 5.2, and we will not dwell on them. As concerns $\forall x A(x)$ or $\exists x A(x)$, things are a little more complex. The point is that we are also dealing with variables here, and an interpretation tells us nothing about them. In order to "interpret" the variables, in some sense, we need to combine a given interpretation with a family of possible *assignments*. These are functions that associate each variable to an individual in D – in other words, an assignment acts as if variables were names. They are usually denoted by $r, r', r'', \ldots$ and so on. We say that two assignments r and r' are x-alternative to one another when they assign the same values to each variable, except possibly x. In other words, if two assignments r and r' are x-alternative (to one another) then they can differ at most in the value they give to *x*, not in those they give to the other variables. This pack of notions is crucial for defining the truth conditions of ∀*x*A(*x*) and ∃*x*A(*x*). Moreover, truth conditions here need to be relativized to the combination of an interpretation and an assignment; we denote combinations of this kind by I*r*. We start from the truth condition of ∀*x*P(*x*), which is:

I*r*(∀*x*P(*x*)) = T if and only if *r*'(*x*) ∈ I(P) for all the assignments *r*' which are *x*-alternative to *r*

Let us try to understand this. First, let us see *where* it is meant to take us, indirectly: it means to tell us that ∀*x*P(*x*) is true if and only if all the individuals in *D* are in the subset *S* of *D* denoted by *P* – that is, if *S* = *D* for *S* = I(P). How does it tell us this? Simple: if we consider all the possible assignments *r*' which are *x*-alternatives to *r*, by considering them together we will find that all the individuals in *D* will be associated, from assignment to assignment, to *x*. The fact that then the assignment of *x* is in the interpretation of *P* for all relevant assignments implies that all the individuals in *D* are in the subset *S* of *D* denoted by *P* according to the interpretation I (which, contrary to the assignment *r*, remains "fixed" in evaluating the truth of the quantified sentence).

Something similar happens with sentences of the form ∃*x*A(*x*). The truth condition of ∃*x*P(*x*) is:

I*r*(∃*x*P(*x*)) = T if and only if *r*'(*x*) ∈ I(P) for some assignment *r*' which is *x*-alternative to *r*

Once again, the condition is meant to tell us that ∃*x*P(*x*) is true if and only if at least one individual in *D* is in the subset *S* of *D* denoted by *P* – that is, if *S* ∩ *D* ≠ ∅ for *S* = I(P). How does it tell us this? Simple: if there is a possible assignment *r*' which is *x*-alternative to *r* and such that *r*'(*x*) ∈ I(P), then we will find that there is at least one individual in *D* which is in the interpretation of *P*.

The two conditions can be easily generalized to sentences of the form ∀*x*A(*x*) and ∃*x*A(*x*), where A is a formula of arbitrary complexity:

I*r*(∀*x*A(*x*)) = T if and only if I*r*'(A(*x*)) = T for all the assignments *r*' which are *x*-alternative to *r*
I*r*(∃*x*A(*x*)) = T if and only if I*r*'(A(*x*)) = T for some assignment *r*' which is *x*-alternative to *r*

Let us now look at the truth conditions of sentences of the form ∀*x*(P(*x*) → Q(*x*)) and ∃*x*(P(*x*) ∧ Q(*x*)), which we have discussed mostly at the beginning of § 5.3. The truth conditions of ∀*x*(P(*x*) → Q(*x*)) are:

I*r*(∀*x*(P(*x*) → Q(*x*))) = T if and only if, for all the assignments *r*' which are *x*-alternative to *r*: if *r*'(*x*) ∈ I(P), then *r*'(*x*) ∈ I(Q)

5 Deductive arguments

Due to the reasoning we have presented above, this ultimately means that $\forall x(P(x) \rightarrow Q(x))$ is true (in a given interpretation I) if and only if $I(P) \subseteq I(Q)$. This justifies the "set-theoretical" reading that we have given to sentences of the form "All Ps are Qs". Similarly, the truth conditions of $\exists x(P(x) \wedge Q(x))$ are:

$I_r(\exists x(P(x) \wedge Q(x)))$ = T if and only if, for some assignment r' which is x-alternative to r: $r'(x) \in I(P)$ and $r'(x) \in I(Q)$

Due to the reasoning we have presented above, this ultimately means that $\exists x(P(x) \wedge Q(x))$ is true (in a given interpretation I) if and only if $I(P) \cap I(Q) \neq \emptyset$. This justifies the "set-theoretical" reading we have given of the sentences of the form "Some Ps are Q".

We close this technical part with two short remarks. First, we cannot represent the truth conditions of sentences of the form $\forall x A(x)$ and $\exists x A(x)$ by truth tables. This is because the quantified constructions are not truth-functional. Second, the semantics for negation from § 5.2 and the semantics of $\exists$ make it clear that we can define the universal quantifier $\forall$ in terms of $\neg$ and the existential quantifier $\exists$. It is easy to see that $\forall x A(x) =_{def} \neg \exists x \neg A(x)$ is an adequate definition. Alternatively, it is possible to define the existential quantifier in terms of the universal one: it is also easy to see that $\exists x A(x) =_{def} \neg \forall x \neg A(x)$ is an adequate definition.

Some logical equivalences

Interpretations are not like the rows of truth tables, but they play the same role: that of constructing the logically possible scenarios that we must consider in order to check the validity of a sentence or piece of reasoning. Here we will use them very indirectly to check not the validity of a reasoning or a sentence, but the logical *equivalence* between different sentences. More specifically, we will use the set-theoretical reading that the interpretations of the quantified sentences permit while checking the truth conditions of these sentences. We will start with:

(1) $\forall x(P(x) \rightarrow \neg Q(x))$ is equivalent to $\forall x(Q(x) \rightarrow \neg P(x))$

The first of the two quantified sentences is true if and only if the set on which P is interpreted is a subset of the *complement* of the set on which Q is interpreted – that is, if the set on which P is interpreted is a subset of "everything that is not Q" – and therefore if $I(P) \subseteq (D - I(Q))$,[17] which is equivalent to $I(P) \cap I(Q) = \emptyset$.[18] The second of

---

17. In set theory, this string of symbols reads as "the set that interprets P is a subset of the complement of the set that interprets Q". The complementation operation – (with respect to a set S which is a subset of a given domain D) is the one that returns all those individuals in D which are "outside" S.
18. In set theory this string of symbols reads as "the set that interprets P and the one that interprets Q have an empty intersection" – which means that they share no members.

the two quantified sentences is true if and only if the set on which Q is interpreted is a subset of the *complement* of the set on which P is interpreted – that is, if the set on which Q is interpreted is a subset "of everything that is not P" – and therefore if $I(Q) \subseteq (D - I(P))$, which is again equivalent to $I(P) \cap I(Q) = \emptyset$. Thus, the two sentences ultimately have the same truth conditions, and are therefore logically equivalent.

That equivalence is also intuitive, and a brief consideration of a couple of sentences in our natural language will help us understand it. Take "No man is immortal". If this statement is true, "No immortal is a man" is also true. If any immortal were a man, in fact, it should be the case that "No man is immortal" is false. In a completely analogous way, it can be understood that if "No immortal is a man" is true, then "No man is immortal" is also true.

The following is a further equivalence:

(2) $\exists x(P(x) \wedge Q(x))$ is equivalent to $\exists x(Q(x) \wedge P(x))$

The first of the two sentences in question is true if and only if $I(P) \cap I(Q) \neq \emptyset$ – that is, if the sets that interpret P and Q, respectively, share at least one member. The second is true under exactly the same conditions.[19]

Once again, let us take an example from natural language: "Some people who live in Milan are football players" cannot be true unless it is also "Some football players (are people who) live in Milan".

Notice instead that:

(3) $\forall x(P(x) \rightarrow Q(x))$ is *not* equivalent to $\forall x(Q(x) \rightarrow P(x))$

This is because $I(P) \subseteq I(Q)$ *does not imply* that $I(Q) \subseteq I(P)$ – in general, $S \subseteq T$ does not imply $T \subseteq S$: S may be a subset of T without T being a subset of S.

For instance, "All men are mortal creatures" is true. But it is not true that "All mortal creatures are men". Furthermore:

(4) $\exists x(P(x) \wedge \neg Q(x))$ is *not* is equivalent to $\exists x(Q(x) \wedge \neg P(x))$

The first sentence is true if and only if $I(P) \cap (D - I(Q)) \neq \emptyset$ – that is: if and only if the set that interprets P and the complement of the set interpreting Q have a non-empty intersection. However, this can also happen because, for example, $I(Q) = \emptyset$ – that is, the set that interprets Q is empty. Under this same condition, however, the second sentence is false, because if it were true, then $\exists xQ(x)$ would also be true. However, this is true if and only if $I(Q) \neq \emptyset$.

For instance, there is at least one man who is not tall and not short at the same time, but this does not imply that there is something tall and short at the same time and that it is not a man. This is because there is nothing that is both tall and short

---

19. This holds because $S \cap T = T \cap S$ for any pair of sets S and T.

5 Deductive arguments

at the same time. Such an object would indeed make a contradiction true, but contradictions are logically false in the logic we assume in this volume. This suffices to show that the two sentences are *not* equivalent.

Let us look at some other equivalences.

(5) $\forall x(P(x) \rightarrow Q(x))$ is equivalent to $\forall x(\neg Q(x) \rightarrow \neg P(x))$

The first of the two quantified sentences is true if and only if the set on which P is interpreted is a subset of the one on which Q is interpreted and therefore if $I(P) \subseteq I(Q)$. It is easy to see that this is equivalent to $(D - I(Q)) \subseteq (D - I(P))$. Indeed, if $(D - I(Q)) \not\subseteq (D - I(P))$, some member of $(D - I(Q))$ is also a member of $I(P)$, which is incompatible with $I(P) \subseteq I(Q)$. This suffices to show the equivalence of the two sentences.

Just to give a concrete example: "All cats are mammals" is true. But this implies that "Anything that is not a mammal is not a cat". It is easy to see that the implication also goes in the other direction.

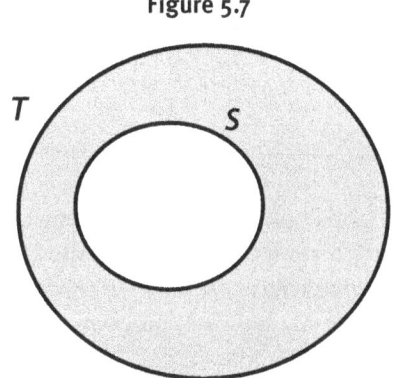

Figure 5.7

Figure 5.7 helps to understand the reasoning behind (5) via the visual representation of the relationships between two sets. All that is in set S is also in T but, as a consequence of this, all that is outside T will also be outside S.

Finally, notice that:

(6) $\exists x(P(x) \land \neg Q(x))$ is equivalent to
     $\exists x(\neg Q(x) \land P(x))$

This actually follows from the commutativity of conjunction.

Introduction and elimination rules for quantified sentences

For the sake of completeness, we will briefly present the introduction and elimination rules for $\forall$ and $\exists$, along the lines of natural deduction for quantified classical logic, i.e. that version of classical logic that can express the kind of sentences that we are considering now.

This is the *rule of universal quantifier introduction*:

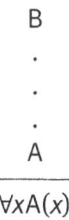

The rule tells us that if I *prove* a formula A, then the formula will express something true for *all* individuals. This is because, in proofs, the choice of one individual (rather than another) is *arbitrary*. In other words, the choice of A, containing, say, names $a_1, ..., a_n$, is arbitrary. What if we choose a sentence B that is the same as A, but for the fact that names $b_1, ..., b_n$ occur in it? Think of the mathematical reasoning where we start from the arbitrary choice of an object ("Take a number n with these given properties", "Take a figure c with these given properties"). What this kind of reasoning does, typically, is to show that n or c also has another property, and finally conclude that, since n or c are arbitrarily chosen (out of numbers or figures that have the given initial property), the result can be generalized to all numbers or figures with the initial property. The scheme of this kind of mathematical reasoning is:

$$[P(a)]$$
$$\vdots$$
$$\frac{Q(a)}{P(a) \rightarrow Q(a)}$$
$$\overline{\forall x(P(x) \rightarrow Q(x))}$$

This is how we read the scheme: If you prove $P(a) \rightarrow Q(a)$, since from the temporary assumption of $P(a)$ it follows that $Q(a)$, then, if a is an arbitrary name, you can also conclude $\forall x(P(x) \rightarrow Q(x))$. The scheme applies the introduction rule of the universal quantifier precisely in inferring the conclusion, together with the rule for introducing the material conditional that we will look at in Chapter 6 (and which is used in the second to last step of the proof).

The *rule of universal quantifier elimination* is:

$$\frac{\forall x A(x)}{A}$$

That is: If you prove $\forall x A(x)$, you can conclude A (where A contains names while $\forall x A(x)$ contains variables). For example, if I prove that each object is red or not red, then I prove that the object denoted by a is either red or not red:

$$\frac{\forall x(P(x) \vee \neg P(x))}{P(a) \vee \neg P(a)}$$

The *rule of existential quantifier introduction is*:

$$\frac{A}{\exists x A(x)}$$

that is, if you assume or prove A, then you can conclude that there is an individual who satisfies A. For example, if I assume that a is red, then I can conclude that there is at least one individual who is red.

The *rule of existential quantifier elimination* is:

$$\frac{\exists x A(x)}{A[x/c]}$$

where c is an individual constant that does not appear in the proof.

That is, if I prove or assume that $\exists x A(x)$, then I can arbitrarily take a "new" individual constant c – that is, a constant that had not already appeared in our reasoning – and I can infer $A(c)$. We can write $A[x/c]$ equivalently, which indicates the result of replacing x with c in $A(x)$.

5.3.3 Some remarks on the logical regimentation of predicates and quantifiers

The "functional" reading of the predicates that we have introduced is insensitive to some differences that are present in the grammar of our natural language. For example, the reading does not distinguish between sentences such as "Some plumbers are chess players" (in which a copula is present), and sentences such as "All horses have four legs" (in which there is no copula). One could think that this "insensitivity" obliterates an aspect of our language that likely plays a role in our reasoning. The issue is actually hard to solve and we will not deal with it in this volume. We will assume, instead, that such insensitivity does no harm to our treatment of reasoning.

A similar point: our formal regimentation does not distinguish between adjectives and generic nouns. It does not see the difference, so to say, between "All hikers are tired" and "All horses are mammals". This insensitivity becomes less significant as soon as we read the two sentences as our regimentation suggests, namely as "All individuals who are hikers are tired" and "All individuals that are horses are mammals".

One last remark before closing these considerations. Here we have considered only some of the possible quantifiers of our language, as is usual when dealing with quantified logic, and we have done this because "all" and "some" are the easiest quantifiers to handle. In any case, the formal regimentation of quantifiers that we have presented applies to other quantifiers, and this has already been done. For instance, we have a formal treatment of what are termed "arithmetic quantifiers" ("There are two/three/four/x's that ..."). Also take sentences like "Only cheetahs can run that fast". We can read it simply using the standard quantified theory we have defined, considering that "Only cheetahs can run that fast" is equivalent to "All animals that can run that fast are cheetahs". Similarly, a statement such as "All but my friends have left" is logically equivalent to the conjunction "No friend of mine has left and everyone who is not my friend has left", which suggests that these two kinds of sentences can be dealt with using the formal means that we have introduced.

5.3.4 Categorical syllogisms

We will now see how the formal regimentation we have introduced helps to analyze a traditional and very famous piece of logic and theory of reasoning: the *categorical syllogism*. This form of reasoning was already studied by Aristotle and then by medieval philosophers and logicians. Categorical syllogisms consist of two premises and a conclusion, each of which is a categorical sentence. In particular, we will see here how the tools we have introduced so far allow us to check the validity or invalidity of a syllogistic inference scheme (or syllogistic scheme). Let us start with an example:

(P1) All dolphins are mammals.
(P2) All mammals are animals.

(C) All dolphins are animals.

In the theory of the syllogism developed by Aristotle and systematized by the logicians of the Middle Ages, the syllogisms are grouped into different types according to the main quantifiers that appear in the premises and in the conclusion. In the above argument, the premises and the conclusion are **A**-sentences, and they can be expressed as follows:

(P1) All *P* are *Q*.
(P2) All *Q* are *R*.

(C) All *P* are *R*.

Medieval logicians gave names with a purely mnemonic function (that is, for the exclusive purpose of remembering what kind of sentences were included in the premises) to the different kinds of syllogisms. The fact that in the scheme indicated above all and only **A**-sentences appear has earned this kind of syllogism the strange "sobriquet" of *Barbara* (the three vowels indeed stand for the kind of sentence appearing in the premises and in the conclusion).

The theory of syllogism of Aristotelian origins also distinguishes the premises on the basis of the predicates that are involved. More precisely, of the two premises, the one of the two premises in which *R* (the predicate of the conclusion) also appears is called the *major premise*. The premise in which *P* (which Aristotle would call the subject of the conclusion) appears is called the *minor premise*. The subject of the conclusion is also called the *minor term* and the predicate of the conclusion *major term*. In general, the major term or the minor term can act both as a subject and as a predicate in the premises in which they appear. In both premises there is a third term *Q*, called the *middle term*, which refers to a third class of objects, different from that to which the major and minor terms refer. The middle term will be the predicate of the minor premise if the minor term appears as subject; it will instead be the subject of this premise if the minor term appears as a predicate. The same is true for the mid-

5 Deductive arguments

dle term and the major premise: it will be the subject or predicate of this premise depending on the function performed by the major term.[20]

Now let us see how the syntactic and semantic regimentation we have introduced helps us to establish the validity of this syllogism:

(P1) $\forall x(P(x) \to Q(x))$
(P2) $\forall x(Q(x) \to R(x))$
―――――――――――――
(C) $\forall x(P(x) \to R(x))$

Note that P1 is true if and only if $I(P) \subseteq I(Q)$ and P2 is true if and only if $I(Q) \subseteq I(R)$. Given the *transitivity* of $\subseteq$,[21] we conclude that $I(P) \subseteq I(R)$. But if $I(P) \subseteq I(R)$, then C is true. Therefore, if P1 and P2 are true, C must also be true. The scheme is therefore *valid* (Figure 5.8).

Another valid syllogistic scheme is the following:

(P1) All *P* are *Q*.
(P2) No *R* is *Q*.
―――――――――
(C) No *P* is *R*.

That is, in formulas:

(P1) $\forall x(P(x) \to Q(x))$
(P2) $\forall x(R(x) \to \neg Q(x))$
―――――――――――――
(C) $\forall x(P(x) \to \neg R(x))$

Notice that P1 is true if and only if $I(P) \subseteq I(Q)$ and P2 is true if and only if $I(R) \subseteq (D - I(Q))$. This means that $I(P) \cap I(R) = \emptyset$ and that, therefore, $I(P) \subseteq (D - I(R))$. But if $I(P) \subseteq (D - I(R))$, then C is true. Therefore, if P1 and P2 are true, C must also be true. The scheme is therefore *valid*. The Figure 5.9 illustrates the relationships between the three sets involved. This scheme of syllogism composed of the

Figure 5.8 *Barbara*   **Figure 5.9** *Camestres*

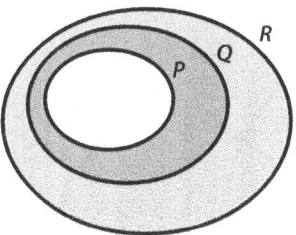
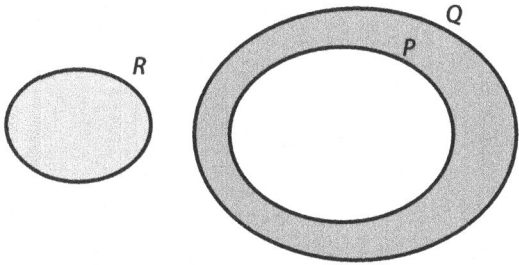

---

20. The minor term is *dolphins*, the major term is *animals*, while the middle term is *mammals*. It is clear why this term is called "middle": the two premises tell us the relations that the two sets denoted by the terms of the conclusion have with that denoted by the middle term.
21. That is: if $S \subseteq T$ and $T \subseteq U$, then $S \subseteq U$.

sequence of type sentences **AEE** – was also given a name in the Middle Ages (*Camestres*).

By contrast, the following scheme is not valid:

(P1) Some P are Q.
(P2) Some R are Q.
―――――――――――
(C) Some P are R.

That is, in formulas:

(P1) $\exists x(P(x) \wedge Q(x))$
(P2) $\exists x(R(x) \wedge Q(x))$
―――――――――――
(C) $\exists x(P(x) \wedge R(x))$

Indeed, P1 is true if and only if $I(P) \cap I(Q) \neq \emptyset$, and P2 is true if and only if $I(Q) \cap I(R) \neq \emptyset$. However, these are compatible with $I(P) \cap I(R) = \emptyset$. For example, suppose that: $I(P) = \{d_1, d_2\}$, $I(Q) = \{d_1, d_3\}$ e $I(R) = \{d_3, d_4\}$ for $d_1, d_2, d_3, d_4 \in D$. It is clear in this case that $I(P) \cap I(Q) \neq \emptyset$, $I(Q) \cap I(R) \neq \emptyset$, and $I(P) \cap I(R) = \emptyset$. See Figure 5.10 for a visual representation of such a situation. Notice that if $I(P) \cap I(R) = \emptyset$, then C is false. The kind of situation just described gives rise to a logically possible scenario in which P1 and P2 are true but C is false.

There are 256 possible forms of syllogism. Indeed, the premises and the conclusion may take the forms **A**, **E**, **I** or **O**. The possible combinations are therefore 4x4x4 = 64. Moreover, in each of the premises the middle term can be in the position of subject or in the position of predicate. So we have 2x2 = 4 further possibilities. The number of possibilities therefore total 64x4 = 256. We can also list which of these 256 possible forms are deductively valid.

Syllogisms are usually classified by figure and mode. There are 4 figures, which depend on the arrangement of the middle term in the premises. In the syllogisms of

Figure 5.10

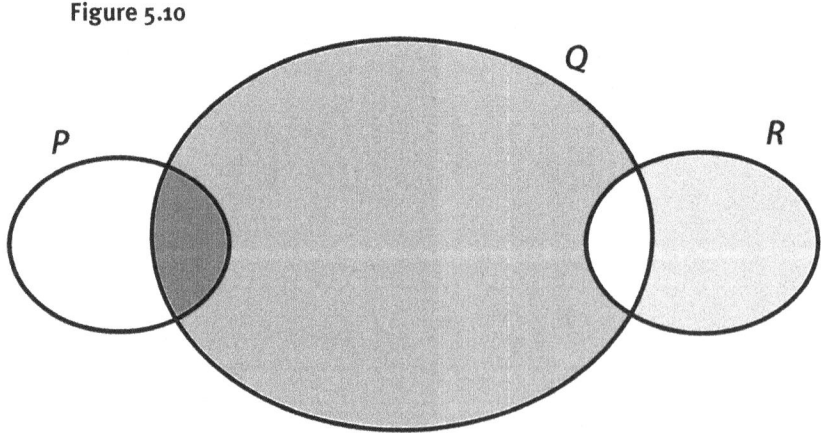

the first figure, the middle appears in the position of subject in the major premise and predicate in the minor one. The *Barbara* scheme, which we considered above, is a scheme belonging to the first figure. In the syllogisms of the second figure the middle term appears in the position of predicate in both premises. The *Camestres* scheme is therefore a scheme of the second figure. If the middle term appears in the position of subject in both premises, the syllogism is said to be of the third figure; if the middle term appears in the position of predicate in the major premise and of subject in the minor one, it is said to be in the fourth figure. The modes, on the other hand, depend on the types of categorical sentences that appear in the premises and in the conclusion. *Barbara* has mode **AAA**, *Camestres* has mode **AEE**. Together, the mode and the figure uniquely identify a syllogism. For instance, we can identify *Barbara* as **AAA**-1 and *Camestres* as **AEE**-2. This gives a general method for compiling a list of all valid forms.

5.4 Conclusion

To sum up, we have distinguished two kinds of reasoning: one in which the sentences contain at most the so-called connectives (negation, conjunction, disjunction, conditional, biconditional), and one in which the so-called quantifiers are also present (among the many possible quantifiers, we have focused on the universal quantifier *all/for every* and the existential quantifier *exists/some*). Furthermore, we have distinguished two kinds of sentences: atomic sentences, which do not contain logical operators, and complex sentences, which contain them. A further distinction introduced in this chapter concerns two angles we can see things from in reasoning and deductive logic: the *semantic* one, which deals with the truth conditions of sentences and the interpretation of names and predicates, and the *syntactic* one, which deals with specifying rules of manipulation for sentences that contain certain logical operators (in our case, connectives or quantifiers). These rules are called *inference rules*. The connection between the two dimensions lies in the fact that we want the rules to be *valid* (and we have seen that the notion of validity is a semantic notion) – in happy cases, such as that of classical logic, we are able to provide rules that allow us to generate all and only valid inference schemes. We then discussed the *truth-functional* semantics of the connectives for negation, conjunction, and disjunction, postponing the discussion of the semantics of the conditional (if ... then) to the next chapter. We have also discussed semantics through the visual representation tool of *truth tables*. The semantics of connectives and truth tables have provided us with a concrete tool to check the validity of a reasoning in which the sentences contain, at most, the connectives we have introduced. We subsequently presented the rules of introduction and elimination of negation, conjunction, and disjunction (in this case too, we have postponed the rules for the conditional to Chapter 6). We then discussed a reading of these rules that takes us out of the narrow and rigorous sphere of mathematical reasoning, for which such rules were first systematized and proposed. We mentioned that these rules help us to generate all valid patterns of propositional reasoning – that is, that reasoning in which sentences are either simple or contain only connectives. We then discussed those forms of reasoning in which predicates and quantifiers are relevant, exemplifying *quantified sentences* and regimenting them in a precise formal language. We proceeded by providing the semantics of these sentences, using the latter to check the validity or invalidity of the arguments involving them. We also mentioned the introduction and elimination rules for quantifiers. Finally, we used the semantics of quantified sentences to check the validity or invalidity of some inference schemes called "syllogistic schemes".

6 | Conditional reasoning, I: The material conditional

In this chapter we will discuss one kind of reasoning with conditionals, namely the one involving the *indicative conditionals*. After exemplifying a concrete case of reasoning that leads to concluding a conditional sentence (§ 6.2), we define conditional sentences and the conditional connective, then we will discuss indicative conditionals and the problem of fixing adequate truth conditions for these (§ 6.3). We will see that one of the many possible answers to the problem is provided by what is termed the material conditional. We will discuss the semantics and syntax of this kind of conditional, as well as some fallacies of reasoning involving it, and the "paradoxes of material implication" (§ 6.4). Finally, we will specify the links between conditionals, on the one hand, and necessary and sufficient conditions, on the other, and we will briefly discuss other views on indicative conditionals (§ 6.5 and § 6.6). We close the chapter by discussing an experiment in the psychology of reasoning that concerns the ability to perform reasoning with a particular inference scheme for indicative conditionals.

6.1 Introduction

One of the most important forms of deductive reasoning involves *conditionals*, that is, sentences of the form "If A, then B". The class of sentences that we call "conditionals" is quite mixed, with the effect that we have many different kinds of deductive reasoning involving conditionals. In our everyday discourse, the two most important families of conditional sentences are those of *indicative conditionals* and *counterfactual conditionals*, respectively.

Indicative conditionals convey the information that if it is the case that A, then it is the case that B. Here is an example:

(1) If Dante never existed, then the Divine Comedy was not written.

Counterfactual conditionals, on the other hand, are used in order to assert that if it were (or had been) the case that A, then it would be (or would have been) the case that B. An example is:

(2) If Dante had not existed, then the Divine Comedy would not have been written.

These two types of conditionals play an important role in our reasoning. For example, thanks to indicative conditionals, we infer what follows from a hypothesis or an assumption. On the other hand, counterfactual conditionals let us imagine a situation that differs from the one we are in and let us determine what would follow if that situation actually occurred. These two kinds of conditionals follow different principles of reasoning. In this chapter we will focus on the kind of deductive reasoning which involves indicative conditionals. We will postpone discussion of counterfactual conditionals to Chapter 7.

The family of indicative conditionals itself is extremely varied. Some kinds of indicative conditionals have been developed in order to construct sentences that are (or would be, in their supporters' opinion) capable of capturing the dynamics of the reasoning that mathematicians deploy in proving their theorems. Furthermore, a moment of reflection helps us see that, in our daily lives, we use indicative conditionals in many different ways. Even if not all these uses have been framed within some normative theory of reasoning, linguistic practice shows that when we talk about "indicative conditionals", we are talking about a very diversified family. For the sake of simplicity, in this chapter we will deal with the "material conditional", a kind of conditional introduced in logic in order to capture the dynamics of proofs as well as those forms of our deductive reasoning that start out from hypotheses. The material conditional has some peculiarities, as we will see from § 6.4 on. The reason why we picked this conditional for our discussion is that the material conditional is the simplest of those which satisfy the rules of inference that we deem desirable for indicative conditionals. We will look at these rules in §§ 6.4.2 and 6.4.3.

The chapter is structured as follows. In § 6.2 we will introduce a reasoning problem involving conditionals that will help us familiarize ourselves with these sentences, introduce the notion of hypothetical deductive reasoning, and look at the role that conditionals play in it. In § 6.3 we will provide rigorous definitions of a conditional sentence and a conditional connective, respectively – with the latter being the main logical operator occurring in a conditional sentence. In addition, we will briefly return to the various kinds of indicative conditionals (§ 6.3.1) and to the problem of truth conditions for conditionals (§ 6.3.2). In § 6.4 we will deal with the material conditional, its semantics and its rules of inference (§§ 6.4.2 and 6.4.3). We will also briefly discuss what are termed *paradoxes of material implication* (§ 6.4.1) and some fallacies of deductive reasoning that concern conditionals (§ 6.4.3). We will then deal with the connection between conditionals and necessary and sufficient conditions (§ 6.5) and discuss other conceptions of indicative conditionals (§ 6.6). Finally, we will discuss an experiment developed by psychologist Peter C. Wason, which shows that applying *modus tollens* proves harder than we might expect for real cognitive agents – such as ourselves.

6.2 A reasoning problem involving conditionals

Take the following problem.[1] Assume there are two people and that:

A. one is a man, the other is a woman;
B. one (let us call him/her *e*) is English; the other (let us call him/her *f*) is French;
C. *e* asserts: "I am a man";
D. *f* asserts: "I am a woman";
E. at least one of them is lying.

Question: Given these assumptions, which of the following sentences must be true?

F. Only the man is lying.
G. Only the woman is lying.
H. The woman is *f* and the man is *e*.
I. The woman is *e* and the man is *f*.

This is not a hard question, and a little informal reasoning suffices to get the right answer. Here, however, we will identify the correct solution of the problem by following a rigorous method: for each of the solutions F, G, H, I, we will assume, for the sake of argument, that the solution is correct – that is: that the corresponding sentence is true, exactly as we assume our starting assumptions to be – and we will see the result of this from a logically rigorous point of view.

Let us start by assuming that F is true – that is: only the man is lying. We have two possible cases:

Case 1: *e* is a man. From F (Only the man is lying) and C (*e* asserts: "I am a man"), it follows that *e* is actually a woman. At the same time, from F and D (*f* asserts: "I am a woman") it follows that also *f* is a woman. But the conclusion that both *e* and *f* are women, contradicts assumption A (one is a man, the other is a woman).

Case 2: *f* is a man. From F (Only the man is lying) and D (*f* asserts: "I am a woman"), it follows that *f* is a man, which does not contradict what we are temporarily supposing, and, from F and C (*e* asserts: "I am a man"), it follows that *e* is a man, too. Both are men, then, but this contradicts A (one is a man, the other is a woman).

Therefore, if we assume that F is true, and that the assumptions A-E are also true, the negation of A must be true. At this point we apply the *reductio ad absurdum* (RAA) and conclude that F is false if A-E are true. Consequently, F is not the correct answer to the initial question.

---

1. The problem is from Paolo Legrenzi, Armando Massarenti, *La buona logica. Imparare a pensare*, Cortina, Milan 2016, pp. 70-72.

The same thing is true for solution G: we apply a reasoning that goes along the very same line as the one we have just applied to F and here too we come to the conclusion that G cannot be the solution to our problem.

Let us then suppose that H is the solution. What follows from it? From H, C and D taken together, the negation of E follows – remember that E is "at least one of the two people are lying" – since both *e* and *f* are telling the truth. In other words, all scenarios where H, C and D are true are scenarios where E is false. Indeed, these are scenarios in which both persons are telling the truth, thus contradicting E. We apply RAA here again, and conclude that H is false if A-E are assumed to be true. Hence, H is not the correct answer to our problem.

Finally, let us suppose that the right answer is I. Obviously, I is compatible with all our starting assumptions: there is at least one logically possible scenario in which A, B, C, D, E and I are all true – that is, the initial assumptions do not imply the negation of I. Let us look at this in some informal way. On the one hand, A and E together imply that one of the two persons is a man, the other is a woman, and that *at least one of them* is lying. On the other hand, I along with C and D additionally tells us that one of the two persons is a man, the other is a woman, and that they are *both* lying. Clearly the two situations are compatible with each other. This suffices to say that I is, among the four options F-I presented, the only solution compatible with assumptions A-E. But we want to understand something more, that is: if A-E is true, so must I (this of course cannot be the case for F-H). To ascertain this, we can proceed as follows: assume that the negation of I is true together with A-E. It is easy to observe that in doing so we arrive at one of the four incorrect answers F-H. Since these are incompatible with A-E (as we have seen, A-E together with each of F-H implies a contradiction), it follows that A-E is incompatible with the negation of I. Let us try to understand why. Assumption E and the negation of I together imply the disjunction of F and G – that is, if E and the negation of I are both true, one of F and G is true. It follows, however, that E and the negation of I together imply the negation of A (see discussion of Case 1 and Case 2), thus contradicting assumption A. Similarly, A-D and the negation of I together imply the negation of E, thus contradicting assumption E.

Hence A-E are incompatible with the negation of I, that is: there is no logically possible scenario in which A-E and the negation of I are true together. Hence, we conclude that in any scenario where A-E are true, I is true. Therefore, I must be true if A-E are too.

Thus, the reasoning we have just exemplified has helped us find the solution to our problem in a rigorous way. But what exactly have we shown with this reasoning? That I is true? *No*. We have proved that *if* A-E are true, *then* so is I. In other words, we have proved the truth of the conditional sentence "If A and B and C and D and E, then I". This is quite different from proving I. In particular, in the previous example, to prove I we would also have to prove (or have proved) that our starting assumptions A-E are true. But we did something different: we just *assumed* those sentences. In other words, we got them on board for reasoning's sake, to see *what logically follows from them*. We have shown that I follows, but since we have not actually committed ourselves to the truth of the sentences A-E (we have only assumed them), we cannot even commit ourselves to the truth of I. However, we can conclude that it is true that "If A and B and C and D and E, then I" – that is, we can commit ourselves to the truth of this conditional sentence.

In other words, if we assume A and prove that B follows from it, then we also prove that the sentence "If A, then B" is true. Whether A is true or B is true are an entirely different business.

This is very important in order to understand one point: the truth of "If A, then B" does not depend on the truth of A – that is: the conditional in question can be true even if A is not – but on the connection between A and B. This connection can be understood as the holding (or not holding) of the fact that "If A is true, then B is true".

Let us sum this up and define a couple of general notions. *Hypothetical deductive reasoning* is that form of deductively valid reasoning that starts from the *assumption* of one or more *hypotheses* (temporary assumptions, made for the sake of argument) $A_1, ..., A_n$ and which allow us to conclude a conditional sentence "If A_1 and ... and A_n, then B", where B is a sentence which, we have shown, necessarily follows from A_1 and ... A_n. Ideally, therefore, hypothetical deductive reasoning consists of two steps: prove B on the basis of the assumption of A_1 and ... A_n, and then take a further step in your reasoning and validly conclude the conditional "If A_1 and ... A_n, then B". The reasoning we have deployed in order to solve the problem above is an instance of this form of reasoning.

The fact that A_1 and ... A_n are (temporary) assumptions means that we have not previously proved them and we are not committing ourselves to their truth. The passage leading from the proof of B on the ground of A_1 and ... A_n to the final conclusion that "If A_1 and ... A_n, then B" has a technical name: "discharging the hypothesis" (or "discharging the assumption"). It is called this precisely because the last passage, whose conclusion is a conditional sentence, makes it clear that we are not committing ourselves to the truth of the assumptions.

6.3 The conditional

A conditional sentence is a sentence of the form

"If A, then B"

From a logical point of view, the expression "If ... then" is a connective. For each pair of sentences A and B we can generate two conditional sentences, obviously: "If A, then B", and "If B, then A".

In logic, the connective "If ... then" is called a "conditional". This term also applies to conditional sentences – which are simply called "conditionals", usually. Here we will do the same: context will always suffice to see whether we are talking about the connective or the sentence.

Following standard logical terminology, given a conditional "If A, then B", we will say that A is the antecedent of the conditional, and B is the consequent of the conditional. Take for instance the following conditional:

(1) If Napoleon is French, then Napoleon is European.

The antecedent of (1) is "Napoleon is French", while its consequent is "Napoleon is European".

A further example is provided by this conditional:

(2) If Gianni has passed the exam, then Maria is very happy.

In this case, the antecedent is "Gianni has passed the exam", while the consequent is "Maria is very happy".

6.3.1 The many kinds of indicative conditional

We have already said in § 6.1 that the family of indicative conditionals is rich and varied. Take for instance:

(3) If the temperature rises, then the ice melts
(4) If the stone hits the window, then the window will break
(5) If Dante never existed, then the Divine Comedy wasn't written.

Sentence (3) expresses a nomological link between two facts, that is, a link determined by a scientific law that relates a certain increase in temperature to the melting of the ice. Sentence (4) expresses a causal link between two facts or events – the impact of the stone against the window causes the window to break. Sentence (5) simply expresses a relation between Dante's existence and the existence of the work called "Divina Commedia". The three sentences express very different ways in which, given a certain condition (the one introduced by the "if"), something else can happen.

We have said already that the material conditional, which we will deal with in § 6.4, has been elaborated in order to capture some crucial aspects of mathematical proofs, completely abstracting from the other kinds of connections that may exist between what acts as a condition and what is to be conditioned. The three examples above, however, help us introduce a crucial problem, to which the material conditional provides an answer (one of many possible ones). This is the problem of the truth conditions of indicative conditionals.

Let us consider (3) and suppose that this sentence is asserted while standing in front of a refrigerating room in which a huge block of ice has formed. Let us also assume that we can increase or decrease the temperature from the outside. What do we do in order to verify or falsify sentence (3)? I increase the temperature, making it go from 0 degrees to 1 degree. If the block of ice does not begin to melt, then the sentence is falsified (notice that the sentence does not specify how much the temperature should rise, only that if it rises, then the ice melts). If, on the other hand, the ice melts, we can say that the sentence is verified. In the first case, we say that conditional (3) is false, in the second we say that it is true. Something similar can be done with (4): we wait for the impact of the stone in flight (suppose that the sentence was asserted while the stone was thrown towards the window). If the window breaks on impact, we will say that the sentence is true. If the window does not break, we will say it is false.

Now take (5). The antecedent is false, since we know that Dante existed. We cannot therefore falsify the conditional, since in order to do this we should have a true antecedent and ascertain that the consequent is false. Notice that in (5) both the antecedent and the consequent are false. How do we evaluate (5), then? Is it true, or is it false? It cannot be falsified, for sure. Well, does this suffice to deem it true, or not? And if we should not deem it true, on what grounds shall we say that it is false?

This helps us introduce two problems. First problem: if the antecedent of a conditional is true and the consequent false, then the conditional is false. But under which *other* conditions is it false? The second: if a conditional has a true antecedent and a true consequent, then it is true. Under which *other* conditions is a conditional true? Let us try to answer these two questions.

6.3.2 Truth and falsity of indicative conditionals

In order to answer the questions above, we must first notice an apparently counterintuitive situation: there are true conditionals that have false antecedents and consequents. Our use of conditionals in ordinary language proves the point. In fact, if I assert (1) or (2), I am not asserting that their respective antecedents and consequents are true: by asserting (2) I am not asserting that Gianni has passed the exam or that Maria is very happy; by asserting (1), likewise, I do not commit myself with either Napoleon being French or with Napoleon being European.

In this respect, asserting a conditional is very different from asserting a conjunction or a disjunction. Indeed, if I assert a conjunction, I commit myself to the truth of both conjuncts, while if I assert a disjunction, I commit myself to defending the truth of at least one of the disjuncts. In the case of conditionals, on the other hand, we do not commit ourselves to the truth of the antecedent or that of the consequent. What we do is commit ourselves to arguing that if the antecedent is true (or we discover it to be true), then the consequent is also true. In other words, asserting C commits me to defending that B is true *if* (it turns out that) A is true.

What I am doing is to exclude the case in which Gianni has passed the exam but Maria is not very happy, and the case in which Napoleon is French but not European. *In other words, I exclude the case (A and not B)*. Having a false antecedent, therefore, does not suffice to make the conditional false. In order to falsify a conditional, it is *necessary* to show that under the hypothesis in which the antecedent is true, the consequent is false.

Now, with (1) we know that the antecedent is true, while when we assert (2) we do not know if the antecedent is true or false. But what exactly happens in situations where the antecedent is false? From what we have seen, this is neither necessary nor sufficient to falsify the conditional. For example, let us look at the following conditionals:

(6) If Napoleon is German, then Napoleon is European.
(7) If Napoleon is German, then Napoleon is Australian.

Are they true, or are they false? Before going further, let us sum up some important points: true antecedent and false consequent are *sufficient* to falsify a conditional, and therefore to evaluate it as false. True antecedent and true consequent are *sufficient* to verify a conditional, and therefore to evaluate it as true. What happens in cases where the antecedent is false? This is neither sufficient nor necessary to falsify the conditional. But does it suffice to verify it?[2]

In general, the examples above show that answering the two following questions is not trivial:

- Question 1: Under what conditions is a conditional false?
- Question 2: Under what conditions is a conditional true?

However, any proper treatment of conditionals must answer them. Otherwise, it would not be able to determine a good theory of deductive reasoning with conditionals. Cases such as those set out in (6) and (7) make the business of answering the two questions hard, because the conditionals they exemplify are somewhat weird. This impression comes from the fact that the exercise of assuming that Napoleon is German seems futile, since we all know that he is French.[3]

---

2. Consider that a false antecedent and a false consequent are not jointly sufficient to falsity a conditional– to see this, just take example (5) from § 6.3.1: intuitively, we would say that that conditional is true, and it has both a false antecedent and a false consequent.
3. As we shall see in the next chapter, the truth of the conditional in question would not be vacuous if it were a counterfactual conditional – this presupposes an attempt at understanding what is the case in situations which differ from the actual one. In this case, however, we would not use indicative conditionals, but counterfactual conditionals such as: "If Napoleon had been German, he would never have waged war on Russia".

There are several logical approaches to indicative conditionals, and all of them provide tools to uniquely determine the truth-value of indicative conditionals, including the "weird" ones such as those in (6) and (7). A very influential logical tradition considers sentences such as (6) and (7) to be *vacuously true*. The reasoning behind this treatment of conditionals is the following: whoever asserts a conditional is saying that, under the hypothesis in which the antecedent is true, the consequent is also true. But he (she) says nothing about what must happen to the consequent if the antecedent is false. Or in other words: the *only* way to falsify a conditional is to show that the antecedent is true and the consequent is false. If the antecedent is false, we have no way of proving that the conditional is false, even if the consequent is false. The impossibility of falsifying it makes the conditional true, albeit in a vacuous way.

The logical tradition we are talking about is that of *classical logic*, which in fact "chooses" the material conditional as a model of the indicative conditional. Indeed, the semantics of the material conditional presupposes the acceptance of the reasoning illustrated above, as we will see below.

6.4 The material conditional

The material conditional is a connective that provides one of the many possible ways to "read" an indicative conditional sentence. We will use the symbol → to denote this kind of conditional connective. Thus, A → B informally reads "If A, then B", and it is a conditional sentence whose (main) connective is a material conditional.

The material conditional is presented as a way of answering the two questions that we have raised in § 6.3.2: "Under what conditions is a conditional false?" and, most importantly, "Under what conditions is a conditional true?"

In particular, the material conditional proposes the following "recipe" to answer both questions, a recipe that establishes the

> *Truth conditions of the conditional.* The conditional A → B is false only if the antecedent A is true and the consequent B is false; it is true otherwise.

What does "true otherwise" mean here? To understand this we must think about the possible combinations of the truth-values that the antecedent and consequent of the conditional can get. Let us look at them in detail.

1. With A – True and B – True: A → B True
2. With A – True and B – False: A → B False
3. With A – False and B – True: A → B True
4. With A – False and B – False: A → B True

The truth conditions of the conditional tell us that *only in the second case is the conditional false*. In all other cases, it is true. If we accept, as we do in this volume, that a sentence can only be true or false, it follows that the following formulation is equivalent to the one we have just provided:

> The conditional A → B is false if and only if the antecedent A is true, and the consequent B is false.

In all other cases, therefore, A → B is true. These other cases include all those in which the antecedent A is false. This leads us to define the following truth table for the material conditional:

Table 6.1 Truth table of the conditional

A	B	A → B
T	T	T
F	T	T
T	F	F
F	F	T

The truth table we have just depicted also shows that the material conditional is a truth-functional connective; that is, the truth-value of A → B entirely depends on the truth-values of A and B (and on their possible combinations). Furthermore, the table also tells us that A → B is logically equivalent to ¬ (A ∧ ¬B) and to ¬A ∨ B.

The intuition that seems to back the material conditional (and its truth conditions) can be reconstructed as the conjunction of the two following assumptions:

1. A conditional is false in all and only those cases in which the conditions under which its assertion is falsified hold good.
2. The only conditions under which the assertion of a conditional is falsified are those in which the antecedent is true but the consequent is false.

This helps us understand why conditional sentences in which the material conditional is the main connective turn out to be true whenever the antecedent is false. A totally different question is whether the material conditional is actually adequate, as it stands, in capturing some plausible form of conditional reasoning. The question is not trivial. Some logicians think the material conditional fails to be adequate – notice that, above, we have rationally reconstructed the intuition that seems to back the material conditional, but this does not mean that we have proved the *plausibility* of 1. and 2.

However, there are two reasons why we are focusing on the material conditional here, rather than other ways of formalizing indicative conditionals. First, the material conditional is the one with the simplest semantics. Second, it provides intuitive inference rules for reasoning with conditionals. We will examine these aspects further in the following subsections.

In any case, when our reasoning involves conditional sentences, and we assume

6 Conditional reasoning, I: The material conditional

the only conditional (connective) in question is the material one, we simply need to remember that the material conditional is a "game" that has its own rules. In order to answer questions on the material conditional, we just have to learn these rules and apply them, even if they do not seem particularly intuitive. This is exactly the same as playing football: you have to remember (among other things and rules) and apply the rule that the goalkeeper cannot grab the ball with their hands if it comes from a back pass from one of their teammates, even if this rule does not seem particularly intuitive to us.

6.4.1 The paradoxes of material implication

The truth table of the material conditional points out that conditionals with false antecedents are all true. This implies that some conditionals will be true even though they do not intuitively "sound" true to us – take, for instance, "If Napoleon was German, then Napoleon was European" and "If Napoleon was Australian, then Napoleon was European": if we play the game of the material conditional, it clearly follows that these two sentences are true. The equivalence between $A \rightarrow B$ and $\neg (A \land \neg B)$ helps us understand why we have a mismatch between our intuitions on indicative conditionals and what the material conditional does. Indeed, it helps us rephrase the two conditionals just mentioned as "It is not the case that (Napoleon was German and Napoleon was not European)" and "It is not the case that (Napoleon was Australian and Napoleon was not European)". Indeed, it is easy to see that the two last sentences are true. Whether a conditional definable in terms of negation and conjunction can really be conceptually sound is a different issue, which we will not discuss in this volume. Consider this example

- (A) $2 + 2 = 5$
- (B) The Earth rotates on itself while tap dancing.

$2 + 2 = 5$ is false – it is a logical falsity or, more precisely, a falsity from the field of arithmetic: it cannot be true. Given the truth table of the material conditional, this suffices to make us conclude that

- (C) $A \rightarrow B$

is true. That is, it is true that "If $2 + 2 = 5$, then the Earth rotates on itself while tap dancing". More generally, if the conditional in question is the material conditional, then whenever A is a logically false sentence, $A \rightarrow B$ cannot but be logically true. Indeed, if A is logically false, then A is false in any logically possible scenario. Given the truth table of the material conditional, this implies that $A \rightarrow B$ is true in every logically possible scenario, regardless of the truth-value of B.

There are even more general considerations to make. For example, if the conditional connective in question is the material conditional, then these two strange inferences are valid:

(8) World War II did not end in 1941. So, if World War II ended in 1941, then gold is an acid.
(9) 2 + 2 = 4. So, if World War II ended in 1941, then 2 + 2 = 4.

These are just two instances of the two following valid inference rules, whose form is:

(*) $$\frac{\neg A}{A \to B}$$

(**) $$\frac{B}{A \to B}$$

The validity of (*) derives from the fact that conditionals with false antecedents are always true. Now if ¬A is true, then A is false. So a conditional like A → B must be true, if so is ¬A. The validity of (**) derives instead from the fact that if a conditional has a true consequent, then it itself is true, no matter whether the antecedent is true or not (as we can check by rows 1 and 2 of our truth table). Hence, any conditional with form A → B is true, if so is B. This explains the validity of (8), whose form is (*), and the validity of (9), whose form is (**).

(9) also helps us understand another intuitively "weird" case: if we are dealing with the material conditional, a conditional sentence whose consequent is logically true is itself logically true.

Conditionals with false antecedents are not the only ones resulting in weird cases. Consider these further examples:

(10) If gold is valuable, then 2 + 2 = 4.
(11) If Napoleon is French, then Christopher Columbus landed in America in 1492.

These conditionals are composed of sentences which are all true. If we interpret the "If... then..." as a material conditional, then the two conditionals are true – see the row of the truth table having T for both the antecedent and the consequent. This notwithstanding, we still feel that considering (10) and (11) true sounds odd. Or rather: it does not seem intuitively correct to consider them true.

We must be careful and distinguish two things, however. We fully understand, based on the truth table above, that if we read (10) and (11) according to the rules of the material conditional, then these two conditionals must be true. However, this does not necessarily mean we should give up our puzzlement. We could indeed insist that considering (10) and (11) true is counterintuitive, and we could then suggest that the material conditional is not adequate to capture our intuitions about conditional sentences, and that it does not account for the way in which we react to a conditional when we are asked to say whether it is true or false.

However, it is important to understand why deeming (10) and (11) true sounds odd to us. The reason is actually very simple: there is no connection between the antecedent and the consequent, in terms of the relevance of the former for the latter. In other words, the consequent is true but the antecedent does not seem to make any contribution to its truth. The fact that gold is valuable is completely irrelevant to the truth of 2 + 2 = 4, and the fact that Napoleon is French is completely irrelevant to the facts concerning Columbus' travels toward America. In short, when we use conditionals in ordinary language, we seem to assume that the truth of the antecedent must be relevant to that of the consequent. In logic, conditionals have been proposed that satisfy this requirement: these are termed "relevant conditionals". Their logical and semantic behavior is quite complex and we will not discuss them in this book. Suffice it to say that, in addition to considering the truth (or falsity) of antecedent and consequent, these conditionals impose a requirement of relevance by virtue of which part of the information provided by the antecedent must also be provided by the consequent. As it is easy to see, sentences (10) and (11) do not satisfy this requirement for any reasonable meaning of the term "information". In fact, they turn out to be false if we read their conditionals as relevant conditionals.

Thus, the material conditional takes a false antecedent and a true consequent to be (each) a sufficient condition for a true conditional. From this, the so-called "paradoxes of material implication" derive. These are a family of sentences which are valid, given the semantics of the material conditional, although saying that they are true sounds completely counterintuitive to us. One of them is:

$$A \rightarrow (B \rightarrow A)$$

It is easy to see that the formula is valid. If A is false, the entire conditional is true because it has a false antecedent; at the same time, if A is true, then $B \rightarrow A$ is true, because it is a conditional with a true consequent. And of course, this makes the whole $A \rightarrow (B \rightarrow A)$ a conditional with a true consequent, and hence a true conditional. In order to see that this is counterintuitive, take the following example:

> If "R" is the initial of my first name, then (if the moon is made of cheese, then "R" is the initial of my first name).

Why does this sound counterintuitive to us? Because B can be chosen *arbitrarily* in the above formula: the formula is valid even if B has no logical connection with A, or more generally no relevance for it (as in our example). Now take a second example:

$$\neg A \rightarrow (A \rightarrow B)$$

Once again: if A is false, (A → B) is true. ¬A → (A → B) is therefore a true conditional, because it has a true consequent. By contrast, if A is true, then ¬A is false, and therefore ¬A → (A → B) is true, since it has a false antecedent. Notice that the formula is closely connected to one of the rules of inference that we have discussed, the one according to which from ¬A we can validly conclude A → B. An example:

> If I didn't go on vacation for Christmas break, then (if I went on vacation for Christmas break, then Santa has moved to a tax haven).

Again, B can be chosen *arbitrarily*, and the choice makes no difference to the validity of the formula. The latter remains valid even when it does not express any logical connection between A and B, or when B is not relevant to A. A closely connected formula is:

$$(A \land \neg A) \to B$$

which is called *Ex contradictione quodlibet*, a Latin expression meaning "anything follows from a contradiction". In classical logic, which is the logic we follow in this volume, and in many other logical formalisms, contradictions are *logically false*, that is, false in every logically possible scenario. But obviously any conditional with a false antecedent is true. If the antecedent is false in every logically possible scenario, the conditional is true in every logically possible scenario, and it is then valid. Here is another example:

> (14) If I invest in bitcoin and I don't invest in bitcoin, then the general theory of relativity is true.

Here, too, there is no connection between "I invest in bitcoin" and "the theory of general relativity is true", which helps us see why the validity of the formula is counterintuitive. Indeed, the formula remains valid regardless of the presence of any logical or relevant connection between its antecedent and its consequent. Similarly,

$$A \to (B \lor \neg B)$$

is valid because so is B ∨ ¬B (B is either true or false, and in each logically possible scenario either B is true, or ¬B is true). A conditional with a valid consequent is itself valid, regardless of the logical connection between antecedent and consequent, or of the relevance of the former to the latter.

We close with one last paradox of the material implication:

$$(A \to B) \lor (B \to A)$$

The formula is valid; indeed, if B is false, then B → A is a true conditional (since it has a false antecedent) and the whole disjunction is true. Furthermore, if B is true, then A → B is a true conditional (because it has a true consequent) and the whole disjunction is true. Once again, however, the validity of this formula does not require any logical or relevant connection between A and B.

Summing this up, the material conditional proves puzzling. Why did we choose it to discuss our reasoning with indicative conditionals, then? We have answered the question already, and now we have the tools to come back to one of the two answers we have offered, namely: the material conditional is semantically simple. When it comes to the semantics of conditionals, nothing is simpler than the truth table that we have introduced for this connective. What we will see shortly will help us to equip ourselves with the necessary tools to understand the second answer, namely: the material conditional is the simplest of the conditionals that satisfy some basic inference rules that we deem desirable for conditional sentences.

Under other methods used to formalize indicative conditionals, just some of the paradoxes of material implication are valid. The relevant conditional we mentioned above, on the other hand, avoids *all* the paradoxes of material implication. Indeed, the logics of relevance, in which this conditional is defined, have been designed in order to avoid such paradoxes. The price to pay for this, however, is a complicated semantics that does not seem to hold much intuitive appeal.

6.4.2 Inference rules with material conditionals

In Chapter 5 we did not present the rules of natural deduction which involve the material conditional. It is now time to remedy this.

The *rule of conditional introduction*, also known as *conditional proof*, is the following:

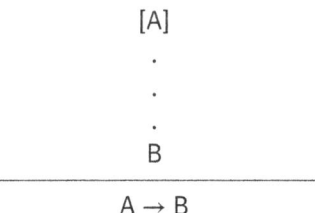

That is, if by starting from the assumption that A we prove B, then we can validly conclude that A → B. In other words, the introduction rule of the conditional is the one we use in hypothetical deductive reasoning. In the example discussed in § 6.1, we just applied the rule in question: once we had shown that sentence I followed from assumptions A-E, we validly concluded that "If A and B and C and D and E, then I". Do not forget that inferring B from A is not the same as asserting A → B. The first is an act that we perform by applying rules of inference or correct proof procedures that lead us from A to conclude B. The second is a speech act that consists in asserting a sentence. Just to have a less abstract understanding of this: in the course of the reasoning that leads from A to B, these two sentences appear, respectively, as the starting premise and the conclusion of a reasoning, while A → B is just one sentence.[4]

The *rule of conditional elimination*, also called *modus ponens,* is one of the crucial rules in deductive reasoning: more specifically, it is one paramount form of deductive reasoning. *Modus ponens* has the following form:

$$A \to B$$
$$\vdots$$
$$\frac{A}{B}$$

The following is an example (an instance) of the rule:

(P1) If Italy imposes medical tests on immigrants, then Italy wishes to ascertain the state of their health.
(P2) Italy imposes medical tests on immigrants.
(C) Italy wishes to ascertain the state of immigrants' health.

It is easy to observe that the rule of *modus ponens* follows from the semantic rule for the conditional: if a conditional is true and the antecedent is also true, then its consequent must also be true. Indeed, by means of (P1) we assert that in the case in which the antecedent is true, the consequent is also true. By means of (P2), however, we assert that the antecedent is indeed true. From this, it follows that if the premises are true, the conclusion must also be true.

6.4.3 Two fallacies and other rules of inference

The inference that we exemplified a few lines above should not be confused with another reasoning scheme which is not valid, and which we now exemplify – a scheme known as the fallacy of asserting the consequent:

---

4. At the same time, however, there is a crucial fact connecting the relation of logical consequence (the fact that a sentence necessarily follows from another given sentence) and the conditional. In particular, if A → B is a logical truth (that is, if that conditional is true in every logically possible scenario), then B necessarily follows from A. In many logics (such as, for instance, classical logic, the one we are following in this book), the other direction holds as well: if B necessarily follows from A, then A → B is a logical truth. This allows us to use logically true conditionals in order to understand the logical connection between the antecedent and the consequent of the sentence (or, as logicians put it, a logically true conditional internalizes in the language the logical connections between its antecedent and its consequent).

(P1) If Italy imposes medical tests on immigrants, then Italy wishes to check the state of their health.
(P2) Italy wishes to check the state of immigrants' health.

(C) Italy imposes medical tests on immigrants.

How can we understand that the reasoning in question is fallacious? Let us consider the second row of the truth table of → (we count by excluding the highest row of all, the one with the sentences). According to it, we see that a conditional with a false antecedent and a true consequent is itself true. Then, coming to our reasoning above, there is a logically possible scenario where (P1) is true, (P2) is true, but (C) is false. Hence, the reasoning scheme that we have just exemplified is not deductively valid.

However, we would not accept this scheme even if we were considering some other indicative conditional instead of the material conditional. Let us see why by using an example that brings us a simpler and more illuminating situation:

(P1) If it is raining, then Gina takes the umbrella.
(P2) Gina takes the umbrella.

(C) It is raining.

Of course, we cannot deductively infer the conclusion that it is raining from the two premises "If it is raining, then Gina takes the umbrella" and "Gina takes the umbrella". For instance, Gina could have, say, taken the umbrella even if it is not raining because she wants to return it to the rightful owner from whom she borrowed it, or because she needs to have it repaired, or for any other reason. So the conclusion does not necessarily follow from the premises: even if the premises are true, it might not be raining at all.

To sum up (generalizing what we have just said), the fallacy of asserting the consequent is the following reasoning pattern:

(P1) A → B
(P2) B

(C) A

This pattern of deductive reasoning is fallacious, that is (i) it is not deductively valid, even if (ii) we tend to follow it for some reason.

There is a further fallacy involving conditionals. It is known as the fallacy of denying the antecedent, and it is exemplified by the following reasoning:

(P1) If we carry out school tests in Italy, then we want to check the work of the teachers in Italy.
(P2) In Italy we do not carry out school tests.

(C) We do not want to check the work of the teachers in Italy.

Again, it is easy to understand why the scheme of denying the antecedent is fallacious. Again, consider the second row in the truth table of → (excluding the one containing the sentences). From it, we see that the logically possible scenario where A is false and B is true is a scenario where A → B is true. There is therefore at least one logically possible scenario in which A → B is true, ¬A is true, and yet ¬B is false. If we replace ¬B by (C), ¬A by (P2) and A → B by (P1), we see that it might be the case that (P1) is true, (P2) is true, and yet (C) is false. As with the fallacy of asserting the consequent, denying the antecedent cannot reasonably be considered as a valid reasoning scheme for any indicative conditional. To understand why, let us apply this pattern of reasoning to the case of Gina and the umbrella:

(P1) If it is raining, then Gina takes the umbrella.
(P2) It is not raining.
―――――――――――――――――――――
(C) Gina does not take the umbrella.

Here too, Gina may have taken the umbrella even if it is not raining, in order to return it to someone from whom she borrowed it. In general, the conditional in question tells us that if it is raining, then Gina takes the umbrella, but it does not exclude that Gina might take the umbrella on other occasions as well – at the same time, it also does not imply that she takes the umbrella on other occasions. It just does not tell us anything about the other occasions Gina might take the umbrella. Thus, we cannot validly conclude, starting from the conditional in (P1) and from the additional information in (P2), that Gina does not take the umbrella. Summarizing (and generalizing), we can say that the scheme of denying the antecedent is as follows:

(P1) A → B
(P2) ¬A
―――――
(C) ¬B

and that this reasoning scheme is fallacious.

We close with a deductively valid reasoning scheme regarding the conditional:

(P1) A → B
(P2) ¬B
―――――
(C) ¬A

This scheme is known as *modus tollens*. In the version of Gina and the umbrella, the correct argument would have this form:

(P1) If it is raining, then Gina takes the umbrella.
(P2) Gina does not take the umbrella.
―――――――――――――――――――――
(C) It is not raining.

It is easy to see that the scheme is valid if the conditional involved is → (the material conditional). Indeed, the fourth row of its truth table (excluding from the count the row containing the sentences) tells us that a true conditional with a false consequent must have a false antecedent. This pattern of reasoning sounds like something that should hold good for any indicative conditional. To understand this point, just keep in mind that the information provided by the conditional is that in the hypothesis that the antecedent is true, the consequent must be true. Now, if the consequent is false, the antecedent must be false if the conditional is to be true, since a true conditional just excludes the fact that the antecedent is true and the consequent false.

Now we have all the tools to appreciate our second answer to the question "why are we considering the material conditional here?" The answer was: because it is the simplest of the conditionals that satisfy the basic inference rules that we require for a conditional. It is the simplest because, of all those conditionals, it is the one that has the simplest semantics. Furthermore, it satisfies all the rules of inference we wish indicative conditionals to satisfy – that is: the ones we saw in §§ 6.4.2 and 6.4.3.

There are specific reasons why these rules are important. Conditional introduction is important because, without it, we could not apply hypothetical deductive reasoning; *modus ponens* is important because without it we could not get the information that the consequent is true from those that the conditional and the antecedent are true; *modus tollens* is important because it codifies the procedure for falsifying a conditional. Indeed, this procedure implies that if a consequent is false and the conditional is true, then the antecedent must be false.

6.5 Conditionals and necessary and sufficient conditions

Reasoning with the material conditional helps us explicitly analyze one of the fundamental elements of informal reasoning, that is: the distinction between *necessary* conditions or properties and *sufficient* conditions or properties. To understand this, let us start with an informal consideration. Consider the following situation:

> *A sufficient condition to pass the exam is that the final grade is equal to or greater than 18/30.*

The condition informs us that only those who achieve a grade equal to or higher than 18 out of 30 pass the exam. Therefore, it is equivalent to stating that:

> *If* you get a grade equal to or higher than 18/30, *then* you pass the exam.

By contrast, consider the sentence that

A necessary condition to take the exam is to register for it.

The sentence tells you that you cannot take the exam unless you have registered for it, that is:

If you want to take the exam, then you must register for it.

Compare this with the previous case. Registering for the exam is a *necessary* condition to take it, but not *sufficient* to pass it, of course, because not all of those who have registered will pass it: to this purpose, it is also necessary to complete the test satisfactorily. On the other hand, getting a mark of 18 or higher in the exam is a *sufficient* condition to pass it, but it is not a *necessary* one to take it – you can take the exam and get a lower mark. We will now give a definition of these notions in more rigorous terms. More precisely, we will use a definition that involves properties. First, we will define the notion of a sufficient condition:

> A property P is a *sufficient condition* for a property Q if and only if: *if* something has the property P, *then* it has the property Q.

For instance, the property

being a dolphin

is a sufficient condition for the property

being a mammal

since the sentence "For all x, if x is a dolphin, then x is a mammal" is true.
We will now define the notion of a necessary condition.

> A property P is a *necessary condition* for a property Q if and only if: everything that fails to have the first property also fails to have the second one.

In our example, therefore, the property

being a mammal

is a necessary condition for the property to

being a dolphin

since the conditional "For any x, if x is not a mammal, then x is not a dolphin" is true. In other words: something is a dolphin *only if* it is a mammal.

Finally, let us come back to the point that something can be a sufficient condition for something else without being a necessary condition and vice versa. A clear example of this:

"Being a father" is a necessary but not sufficient condition for "being a grandfather" because if someone is not a father then he will not be a grandfather either, but someone may well be a father and fail to also be a grandfather (not all fathers are grandfathers, while all grandfathers are fathers).

"Being a father" is a sufficient but not necessary condition for "being a parent" since anyone who is a father is also a parent, but one may fail to be a father and yet be a parent – this is someone who is a mother (all fathers are parents, but not all parents are fathers).

We can easily generalize the notions of sufficient condition and necessary condition to sentences:

A sentence A is a sufficient condition for a sentence B if and only if the truth of A is sufficient to secure the truth of B, that is: if A is true, then so is B.

This definition tells us that A → B indicates that A is a sufficient condition for B. However, the same conditional does not indicate whether A is necessary for B or not. On the other hand, if A → B is true, then B is a necessary condition for A. Again, however, the conditional in question does not indicate whether B is sufficient for A, or not.

These properties immediately relate to the validity of *modus ponens* and *modus tollens*. Since given the truth of A → B, A is a sufficient condition for B, the truth of A is sufficient for the truth of B. This "conceptually" validates *modus ponens* as an inference scheme. On the other hand, if A → B is true, then B is a necessary condition for A. Consequently, if B is false, A must also be. This "conceptually" validates *modus tollens* as an inference scheme.

Now suppose that both A → B and B → A are the case. In that case, A will be a necessary and sufficient condition for B, and B a necessary and sufficient condition for A. If A is true, B must be too, and if B is true, A must also be true; on the other hand, if A is false, B must also be false and if B is false, A must also be false. In logical notation, we shorten the conjunction of A → B and B → A as:

A ↔ B.

The symbol "↔" denotes what logicians call a *biconditional*. It expresses what, in natural language, we express by the phrase "if and only if": "A *if and only if* B" expresses the fact that A and B imply each other. For instance:

Gina takes an umbrella if and only if it is raining

expresses that Gina takes the umbrella if it is raining, and only if it is raining.

The truth table of the biconditional is as follows:

Table 6.2 Truth table of the biconditional

A	B	A ↔ B
T	T	T
F	T	F
T	F	F
F	F	T

If A ↔ B is true, then it cannot be the case that A is true and B false, and it cannot be the case that B is true and A false. In case A ↔ B is true, it must be that either A and B are both true, or A and B are both false.

6.6 Other views of indicative conditionals

As we saw in § 6.4, there are two aspects that make the material conditional hard to accept from an intuitive point of view. First, a false antecedent is sufficient for A → B to be true. Second, A → B is true even if A is completely irrelevant to the truth of B. Consequently, it is understandable that attempts were made to provide alternative views of indicative conditionals, that would prove capable of accounting for the meaning that the connective "if ..., then ..." has in natural language. In § 6.4 we mentioned the "relevant conditional", and many other conditionals have been designed in logic, in order to arrive at a treatment of the "if..., then..." that keeps the connective closer to natural language.

In addition to this, however, something more general can also be done, that is to try to understand what procedure, ideally, leads us to evaluate a conditional as true. In this regard, a viable procedure is the following. When we have to evaluate a conditional, our "evaluation processes" change according to what we think about the antecedent and according to what we already believe, think, hold true in general – that is, according to what our given "stock of beliefs" is. We call this stock of beliefs K. K includes all the sentences we deem true – more intuitively, all the sentences that we believe.

In general, when we face a conditional of the form "If A, then B", what we tend to do can be described as the following procedure, known as the Ramsey Test (named after the British logician, mathematician, economist and philosopher who invented it: Frank Ramsey, 1903-1930):

> (RT) Add A to K and see if, this addition being given, B is true. If this is the case, then "mark" "If A, then B" as true.

The output of a run of RT depends on what we already believe about the antecedent.

6 Conditional reasoning, I: The material conditional

- If we do not already have an opinion on the truth of the antecedent, what we do is a small "thought experiment" – i.e., we add the antecedent of the conditional to our stock of beliefs as a hypothesis. If given this addition, B "follows", then we will also add the consequent B to our stock of beliefs K. This is because rational agents also tend to believe all the consequences of what they already believe, even of what they just believe without being sure of its truth. In this case the "degree of belief" in the conditional, the degree to which we believe in the conditional, will be the same as our "hypothetical belief" in the antecedent of the conditional.
- If we believe that the antecedent is true, then the antecedent is already part of our stock of beliefs. In this case, the degree of belief in the conditional will be the same as the degree of belief we have in the consequent.

Consider an indicative conditional like the following:

> If North Korea ramps up its nuclear program, the United States will intervene militarily.

Imagine that you are a strategic analyst.

- If you want to understand whether this conditional is true, but you do not know whether North Korea will in fact ramp its nuclear program up, then, by applying RT you add the antecedent of the conditional to your stock of beliefs as a hypothesis – which implies that you give it a certain degree of probability (greater than 0, of course). If adding this information to your stock has the "effect" of making you conclude that the US will intervene militarily, then the conditional will be true. This "effect" will depend on the interaction between the hypothesis and some other information in your stock.
- If, on the other hand, you are already certain of the truth of the antecedent, you do not have to make any assumptions: the information is already in your stock. You must therefore "only" assess whether the consequence is true, that is, only whether the US will intervene militarily.

This view of indicative conditionals solves the two problems we have seen above. On the one hand, if A is in contradiction with our stock of beliefs K, that is, if we know that A is false, then it will be impossible to add A to K without running into a contradiction. In this case, we might think that the evaluation process of "If A, then B" stops and it remains indeterminate whether the conditional is true or false. Thus, we will not deem true all the conditionals that have a false antecedent. On the other hand, it is precisely the addition of A to K that makes B true. Hence, A is relevant to the truth of B. Thus, we avoid the counterintuitive phenomenon of deeming true all the conditionals in which A and B are true but in which A is completely irrelevant to the truth of B.

6.7 An experiment involving modus tollens

The Psychology of reasoning has pointed out that we, as real cognitive agents, have a hard time carrying out those forms of reasoning in which we have to start out from the falsity of something. This has been clear at least since psychologist Peter C. Wason's famous experiment back in 1966. The experiment presents a problem in which the application of *modus tollens* is crucial to provide the right solution. The experiment is also called the *Selection task* or *The four-card problem*, and it can be described as follows.

Participants in the experiment are presented with four cards. Each of these has a letter on one side and a number on the other. Two cards are turned onto the letter side (A and B), the other two onto the number side (2, 5), as the figure shows:

Figure 6.1 The four-card problem

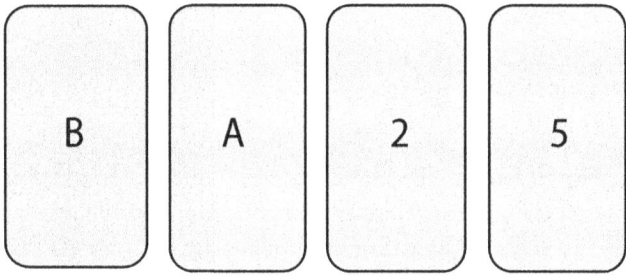

The following *rule* is also presented to test subjects:

> *If there is the letter A on one side of a card, then there is the number 2 on the other side.*

Test subjects are then asked:

> *Tell which cards must be turned in order to determine if the rule is true, or if it is false.*

This is the "selection task" that inspires one of the names of the experiment. In a nutshell, the task is to identify which cards one needs to turn in order to verify whether this rule is true or false. Well, the rule is expressed by a conditional. The correct solution is to flip the card with the A and the card with the 5. Why? Because if the conditional expressed by the rule is true and on a visible side of a card there is A, then on the other side there must be a 2. If on the other side of that card there is another number, then the conditional expressed by the rule is false. Thus, we need to turn the card with the A on the visible side in order to check if the rule is false. Or rather: turning the card with the A on the side may suffice to show that the rule is false. In fact, if there is not a 2 on the other side, we know the rule is false. Notice that the idea of turning the card with the A on the visible side presupposes the ability to reason according to the *modus ponens*. In fact, we pick that move on the ground of the following inference:

(P1) If there is the letter A on one side of the card, then there is the number 2 on the other.
(P2) There is the letter A on one side of the card.

(C) There is the number 2 on the other side.

Since the scheme is deductively valid, if the conclusion is false, then at least one of the two premises must be false. (P2) however is true, because we are considering precisely that card which has the letter A on the visible side. Therefore, if (C) is false, so must be (P1), that is the conditional expressed by the rule.

However, if there is a 2 on the back of that card, *this is not sufficient* to conclude that the conditional expressed by the rule is true. In fact, there may be another card that has an A on the hidden side and a number other than 2 on the visible side. Precisely for this reason, out of the four cards in the example, we must also turn the one showing a 5. In fact, if the letter A is behind it, it follows that the rule is false. Again, this presupposes the ability to reason by applying some inference rule for the conditional. In this case, the rule is *modus tollens*. Indeed, consider the following inference:

(P1) If there is the letter A on one side of the card, then there is the number 2 on the other.
(P2) There is the number 5 on one side of the card.

(C) There is no letter A on the other side.

Once again, since the scheme is deductively valid, if the conclusion is false, at least one of the premises must also be false. Since (P2) is true – we are considering the card with the number 5 – it is (P1) which is wrong. But (P1) is indeed the rule.

Summing this up, what we have to do in the Wason test is to look for all and only the cards that can *falsify* the rule, and the cards that can falsify the rule are:

- The one with the letter A on the visible side;
- The one with the number 5.

Instead, we do not need to turn the card with the number 2 on the visible side: the rule is perfectly compatible with the fact that some cards with the 2 have a different letter from the A on the other side. In fact, the rule says that having the letter A on one side is a sufficient condition for having the number 2 on the other, but not that it is a *necessary* condition. Thus, turning the card with the number 2 on the visible side does not help us falsify the rule, and is therefore not relevant. There is no need to flip the card with the letter B either, because the presence of any number behind it is irrelevant to whether the rule is true or not.

The results of Wason's experiment revealed that a very high percentage of subjects correctly state that the card with the letter A on the visible side must be turned, but of these, very few add that it is also necessary to turn the card with the number 5 on the visible side. What makes the experiment hard to get right? Not the part relating to the application of *modus ponens*, given the high percentage of people who identify the card with the A on the visible side as a card to be turned. What we struggle with is thinking about the *other* possibility that *falsifies* our rule, that is, the one that involves reasoning by *modus tollens*. Advocates of so-called *natural mental logic*[5] read the results of this experiment as a confirmation of this thesis: the majority of people fail to have *modus tollens* in their "natural" repertoire of inference rules. This is why the problem strikes so hard.

In any case, even if one does not share this dramatic view on the issue, we cannot deny that the experiment suggests that we, as real cognitive agents, have trouble reasoning by *modus tollens* or, perhaps more appropriately, that *modus tollens* is for us a cognitively complex and unnatural pattern of reasoning.

Here are the statistics of the replies to the selection task – we have only included the two replies with higher average percentages:

- choice of A and 2: 60 – 75%
- choice of A and 5: 5 – 15%

Wason's experiment is very interesting when it comes to the relation between the normative dimension (and normative approaches) and the descriptive dimension (and descriptive approaches) in reasoning. The results of the experiment tell us that, on average, we cannot easily reason according to *modus tollens*. In short, *modus tollens* does not describe a scheme that we can observe in our reasoning as real cognitive agents. Should we then stop taking it into consideration and proposing it as an inference scheme for our reasoning with conditionals? Not at all. Even if it is a pattern that we have a hard time applying, it is a form of reasoning that we need. Indeed, only by applying it can we correctly solve Wason's selection task: if we do not turn the card with the number 5 onto its visible side, we will not really be able to tell whether the rule is true or false, since we may have left a case which could falsify it unexplored. Taking an example from real life, we should consider that the alibi asserted in a criminal case works on the basis of *modus tollens*: by showing that the accused was elsewhere, the accusation against him is falsified. More generally, we need *modus tollens*, because it expresses a recipe for the falsification of the con-

---

5. For instance, cf. Barbara Rumain, Jeffrey Connell, Martin D. Braine, "Conversational comprehension processes are responsible for reasoning fallacies in children as well as adults: If not the biconditional", *Developmental Psychology*, 19(4), 1983, 471-481.

ditional. The fact that we do not master it naturally, therefore, does not mean that we could do without it. In short, the normative dimension proves to be important for our reasoning.

6.8 Conclusion

In this chapter, we have discussed the form of deductive reasoning that involves the indicative conditional. We have exemplified hypothetical deductive reasoning, and we have examined the role the indicative conditional plays in it. We have briefly discussed the variety of indicative conditionals that we encounter in our use of language and in the formalization of mathematical reasoning, while giving some examples. We have discussed the problem of defining truth conditions for the conditional, and introduced the material conditional, which provides one answer (among many other possible ones) to the problem. We have discussed its semantics and the valid inference rules connected with it, as well as some fallacies and the "paradoxes of material implication". We have looked at the connections between indicative conditionals and necessary and sufficient conditions, and some other views of the indicative conditionals. Finally, we have discussed an experiment in the Psychology of reasoning which highlights some of our problems when applying one of the inference rules that are valid for the material conditional, namely: *modus tollens*.

7 | Conditional reasoning, II: The counterfactual conditional

In this chapter we will discuss another kind of reasoning with conditionals, namely the one involving the *counterfactual conditionals*. We will see that the semantics of these conditionals is different from that of material conditionals and that, more generally, they must be kept distinct from indicative conditionals. In addition to dealing with the logic of counterfactuals and some fallacies, we will discuss some experiments in the psychology of reasoning designed to understand how we reason with counterfactuals. At the end of the chapter we will briefly touch on the links between counterfactual reasoning and causal reasoning.

7.1 Introduction

On 12 May 2013, an article entitled "Sennheiser is leading a campaign against fake electronic goods" appeared in the pages of the *Financial Times*. Sennheiser Electronic GmbH & Co. KG is a major manufacturer of headphones and earphones, and the article announced its campaign to combat the spread of counterfeit products. According to the company, the market for fake Sennheiser products, in addition to defrauding consumers, would lead to a substantial increase in unemployment. According to the CEO's statement, it would cost the company $ 2 million in lost income. To judge the extent of the loss, you need to evaluate how much you would earn *if there were no* counterfeits market. Only by imagining this situation, and the consequences deriving from it, can it be argued that this market generates an economic loss, and how huge this loss is. In short, the reasoning of the CEO of Sennheiser involves the ability to imagine *scenarios contrary to the facts* (such as the one in which the counterfeit headphones market does not exist). From now on, we will call the latter "counterfactual scenarios". These arguments contain pitfalls: in order to think about how much one would have earned if the counterfeit products market did not exist, it is necessary to apply an evaluation criterion. The problem is that there are many alternative criteria. Let us see why.

The CEO has declared $ 2 million as the value of the loss for Sennheiser because this is the multiplication of the sales volumes of *fake* Sennheiser products by the average price of *real* Sennheiser headphones (which amounts to $ 300 per headphone). More precisely, the CEO's reasoning can be captured by the following sentence:

(1) *If there were no* fake Sennheiser products, everyone who bought the fakes *would have bought* real Sennheiser products.

The criterion presupposed by this kind of reasoning can be summarized as follows: To know whether B would have happened if A were not (or had not been) true, imagine a counterfactual scenario where A is false and *almost everything else* is the *same* as it is in the real world. Then ask yourself: Would B be true or false in that scenario? If B is true, sentence

(2) If A were the case, then B would be the case

is true. Otherwise, it is false. Let us see how this actually applies to the CEO's reasoning. We want to figure out whether in the event that A ("There are fake Sennheiser products") *were false*, B ("Everyone who bought Sennheiser fakes buys real Sennheiser products") *would be true*. To do this, let us imagine a scenario where A is false, but almost everything else is the same as in the real world. By "almost everything else", we mean that we do not take any decision on the truth (or falsehood) of B in the counterfactual scenario. What happens in this scenario? The persons who *in the real world* have bought a counterfeit product *want to* buy a product that looks completely like Sennheiser real products – the same desire they have in the real world. In the counterfactual scenario, however, the only products that appear to be Sennheiser in all respects are the *real* Sennheiser products, because in such a scenario there are no counterfeit products. Nobody forbids people from buying Sennheiser products in the counterfactual scenario – just like in the real world. Hence, it is natural to conclude that B is *true* in the counterfactual scenario. *If* we follow the criterion in question, then, we conclude that sentence (1) is true, and we ideally reason as we have imagined the CEO of Sennheiser reasons.

Reasonings based on this criterion are very frequent in estimating losses such as those we have discussed. The cost to the music and film industry of illegally downloading music and films is estimated to be in the billions of dollars, the university book copying market costs the publishing industry millions of euros, etc. In other words: the amount that goes into the counterfeit market, or photocopies, is "added" to the income of producers, and this procedure is ideally captured by the reasoning we have set out. There is, however, something deeply wrong with this reasoning criterion.

In particular, it is not correct (or at least not *always* correct) to reason by "changing" the truth-value of sentence A and leaving almost everything else the *same*. Changing the truth-value of A can, in fact, lead us to review other aspects of reality, and we cannot establish *a priori* which aspects need to be changed. In the case we have just considered, we cannot simply "cancel" the counterfeit market. The absence of this market may lead some of those who bought the counterfeit goods to buy no product. We cannot therefore think that *everyone* who really wanted Sennheiser products and bought counterfeit products would have bought Sennheiser products if there were no counterfeits. Again, the counterfactual scenario by which we modeled the CEO's reasoning is less *close* (less "similar") to reality than another, i.e. the scenario in which (i) the counterfeit market no longer exists, and yet (ii) some of those who bought fake Sennheiser headphones judge that real Sennheiser products are too expensive for them (two real facts are kept fixed: the price of real Sennheiser products and that the price of a good is often negatively correlated with demand).

Since this second counterfactual scenario is *closer to* or more *similar to* the real world than the first counterfactual scenario we have described, we should use the second scenario, and not the first one, to evaluate the truth-value of (1). This is because, in evaluating a counterfactual, we consider scenarios in which the antecedent is true, but which differ *as little as possible* from the real world. Obviously, if we do that, (1) becomes false, not true as in the reasoning we imagine the CEO of Sennheiser did.

Let us consider in more detail what the distinctive features of counterfactuals are. Sentence (1) is a *counterfactual conditional*, that is, a conditional that states *what would be the case* if *something else were the case*. The form of such sentences is:

If A *were the case*, then B *would be the case*.

Or:

If A *had been the case*, then B *would have been the case*.

As in indicative conditionals, A is the *antecedent* and B the *consequent*. Such conditionals are usually asserted when the speaker knows, or believes, or in any case assumes, that the antecedent is *false* in the real world. In contrast to indicative conditionals, counterfactuals are not used to talk about the real world, but about scenarios *alternative* to the real world – what we have defined as "counterfactual scenarios". This is enough to see that counterfactual conditionals cannot be considered false under the same conditions as indicative conditionals. Since we usually assume that the antecedent of a counterfactual is false, if we adopted the truth table of material conditionals seen in the previous chapter, each counterfactual would be true.[1] Yet, we consider some of them to be false. It would not make sense to consider them true under the same conditions as indicative conditionals, because counterfactuals are designed to have an antecedent that is not true. In this chapter, among other things, we will be concerned with understanding under what conditions counterfactuals are true or false. To be more precise: we will give a semantics for

---

1. To avoid misunderstanding and to distinguish the kind of conditionals we are dealing with in this chapter from indicative conditionals, we will call them *counterfactual conditionals* or simply *counterfactuals*.

counterfactuals, and thus we will look at the general criterion on the basis of which we judge that a counterfactual is true or false.

The chapter proceeds as follows. In § 7.2 we will consider some elements of the psychology of counterfactual reasoning. In particular, we will discuss how real cognitive agents tend to reason with counterfactuals. This implies that we will briefly go through a *descriptive approach* to counterfactual reasoning. In § 7.3 we will discuss the logic of counterfactuals, with particular attention to their semantics and to some fallacies we tend to commit, projecting valid reasoning schemes for indicative conditionals onto counterfactuals. Finally, in § 7.4 we will discuss the role played by counterfactual reasoning in the assessment of causal links.

A brief observation before we start: there are two important reasons why it is appropriate to discuss the psychology of counterfactual reasoning, even in a book whose perspective is mostly normative. First, reasoning with counterfactuals has received a rigorous logical and normative approach only in the last forty or so years. This is also due to the fact that this kind of reasoning is one of the most difficult to understand: contrary to other kinds of reasoning, our actual linguistic practice with counterfactuals is so complex and varied that it is difficult to deal with them systematically. Even if we have such an approach today, and even if it is very rigorous and clear, it is appropriate to start out from how we actually reason with counterfactuals to understand the difficulty to deal with them. Second, counterfactual reasoning is important in several relevant practical contexts such as decision-making and strategic interaction – and therefore, indirectly, for disciplines such as Decision Theory, Game Theory, and Economics (see Chapter 1 for these disciplines). As we have seen, if we want these disciplines to provide us with realistic models for predicting the behavior of agents, then we need to understand how agents *actually* reason. This leads us to consider the psychology of counterfactual reasoning.

7.2 The psychology of counterfactual reasoning

The example we discussed at § 7.1 shows that counterfactual reasoning is involved in practical situations of some importance. Here, among many other things, we will briefly list a few important activities in which we are often engaged and in which counterfactual reasoning is fundamental. We will then see in § 7.3 that there are precise ways to discriminate between correct and incorrect counterfactual reasoning.

7.2.1 The importance of counterfactual reasoning

We are very often engaged in cognitive activities that presuppose the ability to evaluate possibilities alternative to the current situation. For some psychologists, such as Philip Johnson-Laird, the ability to represent possibilities (which he calls "mental models") is one of the characterizing traits of our rationality.[2]

In the last thirty years of research in cognitive psychology, two theses have emerged on which the attention of scholars has focused:

- In many cognitive activities, we tend to represent alternatives to reality by modifying *some* aspects rather than others. These alternatives are what we have called "counterfactual scenarios", and this ability is called "counterfactual imagination".
- Counterfactual imagination is "rational" in the sense that it is guided by the same principles that guide rational thinking.

The tendency to construct counterfactual scenarios seems to be presupposed by several common cognitive activities:

- *Learning from mistakes*: if I have made a mistake in a real situation s, I learn from the mistake only if I imagine a situation s' alternative to s that differs from s only in the fact that I do not perform the action that led me to make the mistake and the consequences of that mistake do not occur.
- *Making decisions*: If in a situation s I must decide whether to take action a or not, I imagine a situation s' that differs from s only in the fact that in s' I do a and I compare it with a situation s'' that differs from s only in the fact that in s'' I do not do a. My decision will be guided by my assessment of which alternative produces the best consequences.

The tendency to construct *counterfactual alternatives* (another name for counterfactual scenarios) is also presupposed by some emotional reactions:

- *Having remorse*: usually one feels remorse for *having performed an action* in a situation s; this emotional reaction can be considered as the response to the representation of a situation s' alternative to s in which one has not carried out the offending action and in which one assesses that things would have been better.
- *Having regrets*: usually one regrets *not having performed an action* in a situation s; this emotional reaction can be considered as the response to the representation of a situation s' alternative to s in which one has performed a certain action and in which one assesses that things would have been better.

7.2.2 The fault lines: which aspects of reality do we modify?

The phenomenon of *fault lines* is one of the aspects that most interested scholars of counterfactual phycology. When we mentally represent a certain real scenario, we tend to consider *some* aspects *rather* than others *counterfactually modifiable*.

---

2. Philip N. Johnson-Laird, *Mental Models*, Cambridge University Press, Cambridge 1983.

Psychologists call these aspects *fault lines of reality* and several studies in cognitive psychology have inquired which *fault lines* we change most frequently when we picture counterfactual alternatives.

Here are some examples:

- *Exceptional events*: we imagine counterfactual alternatives to events that we consider "exceptional" or "unusual" rather than to events that we consider "normal".
- *Actions*: we more frequently imagine counterfactual alternatives to actions we have taken rather than to actions we have not taken (that is, we tend to feel remorse more frequently than we have regrets).
- *Controllable events*: we envision counterfactual alternatives in which we modify an event that we believe is under our control rather than an event that we do not consider under our control.
- *Obligations*: we imagine counterfactual alternatives in which we do what we should have done but we have not done.
- *Time*: we tend to imagine counterfactual alternatives to events that occur later in a certain temporal series rather than to events that occur before.
- *Causes*: we envision counterfactual alternatives when we are looking for causal links.

We present a famous experiment carried out by Daniel Kahneman and Amos Tversky, the two scholars already mentioned in Chapter 1. The experiment exemplifies the points we have just considered. It is described in a work entitled *The simulation heuristic*, published in 1982.[3] In this experiment, a group of 62 college students from the University of Pennsylvania were offered the following story (defined by Kahneman and Tversky as an "emotional script"):

> *The Story of Mr. Jones.* Mr. Jones is 47 years old, has a wife and three children and is a successful bank executive. His wife has been home sick for several months. On the day of the accident, Mr. Jones left his office at the usual time. Sometimes he gets off work earlier to do housework at his wife's request, but not on that occasion. Mr. Jones did not go home by the usual route. It was a beautiful day and Mr. Jones told his friends that he wanted to go home by the coastal road to enjoy the view. The accident occurred at a large intersection. The light turned yellow just as Mr. Jones was approaching the intersection. Witnesses stated that he abruptly braked to stop, although he had plenty of time to pass. His family stated that this was Mr. Jones' usual driving behavior. When the light turned green again and Mr. Jones began to cross the intersection, a truck crossed

---

3. In Daniel Kahneman, Paul Slovic, Amos Tversky (eds), *Judgment under Uncertainty. Heuristics and Biases*, Cambridge University Press, Cambridge 1982, pp. 201-208.

with the red light at great speed hitting Mr. Jones's car on the left side. Mr. Jones died instantly. Investigations revealed that a very young boy was driving the truck under the influence of drugs.

The wealth of detail in the story is intended to generate a feeling of "empathy" towards Mr. Jones's family. The students were then asked the following question:

Question: As is commonly the case in these situations, in the days following the accident Mr. Jones's family and friends often thought and said, "If only ...". How did their thoughts or sentences continue? Write one or more likely continuations.

Obviously, what relatives and friends thought was "If only X had / hadn't happened, now Jones would be alive". Kahneman and Tversky wanted to know what real cognitive agents substitute for X.

To answer this question, subjects must imagine a counterfactual scenario (a) to the fragment of the world described in the story, in which (b) Mr. Jones does not die. This can be done only by changing at least one aspect of the story. The answers were organized, roughly, around four counterfactual alternatives. The results of the experiment were the following:

- alternative situations in which Mr. Jones goes the *usual way*: 53%;
- alternative situations in which Mr. Jones *crosses the intersection* with the yellow light: 22%;
- alternative situations in which *there is no young drug-addicted driver*: 20%;
- alternative situations where Mr. Jones leaves the office at a *different time*: 3%;
- other: 2%.

To another group of students, Kahneman and Tversky proposed a small variation on Mr. Jones's story that contained this difference:

Mr. Jones: time variant. On the day of the accident, Mr. Jones left the office earlier than expected to do some household chores at his wife's request. He went home by his usual route. Occasionally, Mr. Jones chose to drive along the coast to enjoy the view on particularly clear days, but that day was just average.

The distribution of the results for this version of the story was as follows:

- alternative situations in which Mr. Jones *crosses* the intersection with the yellow light: 31%;
- alternative situations in which *there is no young drugged driver*: 29%;
- alternative situations in which Mr. Jones returns home at the *usual time*: 26%;
- alternative situations in which Mr. Jones returns home by *another route*: 13%;
- other: 1%.

These distributions of results led Kahneman and Tversky to propose the following theses:

- Most of the subjects tends to modify an event considered exceptional or "anomalous" (the fact that Mr. Jones decides to take the coastal road on the day of the accident). According to Kahneman and Tversky, this demonstrates that subjects prefer the situations they call "downhill", i.e. those in which exceptional events are modified.
- None of the subjects involved in the experiment eliminated any "normal" aspect of the real situation, even those aspects which are very improbable. For example, no subject involved in the experiment has imagined counterfactual alternatives in which Mr. Jones arrives at the intersection 2 or 3 seconds earlier and crosses it with the green light. The incredible coincidence that causes the crash in the story is a very unlikely event. For Kahneman and Tversky, this shows that often we do not tend to imagine alternatives to the least probable events.
- None of the subjects involved in the experiment imagined a counterfactual alternative in which an unlikely event occurred (a type of alternative that Kahneman and Tversky refer to as "uphill"). For example, no one imagined a counterfactual alternative in which the coastal road is blocked by a landslide before Mr. Jones passes by.

We have seen how in fact we tend to reason when we construct counterfactual scenarios. This is also interesting in connection with what we will say shortly, because the ideas on which the semantics of counterfactuals is based seem to conform precisely to some aspects of our counterfactual reasoning highlighted by Kahneman and Tversky's experiments.

7.3 The logic of counterfactuals

In §§ 7.1 and 7.2 we have looked at various examples of counterfactuals. The following further examples will help us distinguish two broad types of counterfactuals:

(3) If employment hit a certain percentage, inflation would rise;
(4) If kangaroos did not have tails, they would fall;
(5) If I hadn't come to Critical Thinking class today, I would have had more free time;
(6) If Stauffenberg had placed a bomb on the right side of the table on 20 July 1944, Hitler would have died.

It is easy to observe that counterfactuals (3) and (4) have the following form:

If it *were the case* that A, then it *would be the case* that B.

These counterfactuals say that the (non-)occurrence of a certain state of affairs or fact (expressed by the antecedent) has a consequence (expressed in the consequent). Sentences (5) and (6) have instead this form:

If it *had been the case that* A, then it *would have been the case that* B.

These counterfactuals say that the (non-)occurrence of a certain *past* state of affairs or fact (expressed by the antecedent) has a consequence (expressed in the consequent). In what follows, however, we will abstract from the distinction between these two types of conditionals.

We will introduce a symbolic notation to be able to discuss the formal treatment of counterfactuals. First, let us note that in natural language it is the *mood of verbs* that indicates that we are dealing with a counterfactual and not with an indicative conditional. For example, in

If Dante *had died* at the age of five, we *would not have* the Divine Comedy

"had died" and "we would not have" tell us that we are talking, respectively, of something that did not happen, and of its supposed consequence. In "If Dante died at the age of five, then we don't have the Divine Comedy", the indicative mood indicates instead that we are talking about the real world. In other words, it is the mood of the verb that takes charge of expressing the "counterfactuality" (or "indicativity"). The formal treatments of counterfactuals, however, do something different: it is the kind of *conditional connective* involved in the sentence that indicates the kind of conditional. This implies that in such treatments the "If... then" of a counterfactual conditional is rendered by a connective different from those for indicative conditionals (for example, the material conditional). Let us see what exactly this distinction between natural and formal language counterfactuals amounts to. In this book, we will use the connective $\Box\!\!\rightarrow$ to express the counterfactual connective:

A $\Box\!\!\rightarrow$ B is to be read: "If it were the case that A, it would be the case that B".

Recall that A and B are schematic letters and stand for the same kind of sentences expressed in the indicative conditionals we have seen so far. So, A could stand for "Dante died at five" and B for "We don't have the Divine Comedy". Consequently, A $\Box\!\!\rightarrow$ B is read as "If it were the case that 'Dante died at five', then it would be the case that 'We don't have the Divine Comedy'". Or, more naturally: "If Dante had died at five, we would not have the Divine Comedy".

Therefore, in formal treatments, the connective $\Box\!\!\rightarrow$, and not the sentences A and B, signals the fact that we are dealing with a counterfactual – that is, with a sentence that speaks of what would have happened if something else had happened. These considerations give us a clear idea of how the *syntax* of counterfactuals works in formal approaches – that is, of how such approaches *construct* counterfactual sentences.

7.3.1 Indicative and counterfactual conditionals

Counterfactuals are different from the indicative conditionals discussed in Chapter 6. To illustrate this, let us take the following pair of conditionals:

(7) If L.H. Oswald didn't kill Kennedy, someone else did;
(8) If L.H. Oswald hadn't killed Kennedy, someone else would have.

The former is an indicative conditional, while the latter is a counterfactual conditional. To understand the context in which we will discuss these two sentences, recall that Lee Harvey Oswald was identified in the so-called Warren report as the assassin of US President John F. Kennedy in 1963, and that, according to this report, he acted alone and not as part of an organized plan or conspiracy. All the hypotheses that contradict these conclusions of the Warren report were based on evidence which was later disproved, and therefore we will here assume that the conclusions of the Warren report are true.

If we abstract from the mood of the verbs in (7) and (8), both of these conditionals have the same antecedent ("L.H. Oswald didn't kill Kennedy") and the same consequent ("Someone else does it"). Suppose that someone makes the hypothesis that counterfactuals and indicative conditionals are the same type of conditionals. How can we prove that this is not the case? The following two facts prove that this hypothesis is wrong:

- (7) is *true*: since John F. Kennedy was really murdered, either L.H. Oswald or someone else murdered him;
- Instead (8) is *false*: Oswald acted alone and there was no collective plan to assassinate Kennedy, if we assume the correctness of the Warren report. So, if Oswald hadn't killed Kennedy, no one would have.

Since (7) and (8) contain the same sentences, however, if they expressed the same kind of conditional, they should be either both true or both false. But this is not the case.

Another way to arrive at the same conclusion is the following. Suppose someone shared the conclusions of the Warren report. They would give two different answers to the questions:

1. What happened if it wasn't Oswald who killed Kennedy?
2. What would have happened if Oswald hadn't killed Kennedy?

They would answer the first question with: "Someone else killed him", and the second with: "No one else would have killed him". These two questions require you to give variable X a value that makes the following sentences true: "If Oswald did not kill Kennedy, then X" and "If it were the case that Oswald did not kill Kennedy, then it would have been the case that X". Since the two conditional sentences are formed by the same elementary sentences, the fact that the speaker gives two incompatible values to the Xs ("Someone else killed him", "No one else would have killed him") indicates that (7) and (8) express two different kinds of conditional.

7.3.2 The truth conditions of counterfactuals

We saw in § 7.1 that we cannot give counterfactuals the same truth conditions we give to material conditionals (and, more generally, to indicative conditionals). In fact, only if an indicative conditional has a true antecedent and a false consequent, is it false. Counterfactuals, however, generally have false antecedents, and therefore it makes no sense to keep these truth conditions. Having generally no true antecedent, they could never be falsified. Yet, there are false counterfactuals, of course: a glance at sentence (10) below is enough to remove any doubts in this regard.

Another important point is that the truth conditions of material conditionals cannot be applied to counterfactuals. For a material conditional to be true it is sufficient that the antecedent is false. But since we almost only assert counterfactuals with false antecedents, we should conclude that we almost only assert true counterfactuals, if the rules of indicative conditionals are extended to counterfactuals. But there are false counterfactuals, such as sentence (10) below.

These two considerations make it clear that counterfactuals need their own semantics. This semantics should allow us to answer the following questions:

- Question 1: Under what conditions is a counterfactual true?
- Question 2: Under what conditions is a counterfactual false?

We will answer these questions in § 7.3.4. Before doing so, we will introduce a test that provides a criterion for accepting (as true) or rejecting (as false) a counterfactual. It is derived from the Ramsey Test for indicative conditionals and to distinguish it from the latter we will call it the *Counterfactual Ramsey Test*. Obviously, "accepted as true" and "true" are not the same thing – and the same goes for "rejected as false" and "false" – but the Test is useful to understand some aspects of the semantics that we will provide for counterfactuals (in particular, some semantic moves are similar to those of the Ramsey Test).

In any case, the semantics of counterfactuals *cannot* be *truth-functional*, that is: the truth conditions of sentences of the form A $\mathbin{\Box\!\!\rightarrow}$ B cannot be determined entirely by the truth-values of A and B. To realize this, it is sufficient to consider the following two sentences:

(9) If Dante had died at the age of five, then we would not have the Divine Comedy;
(10) If Dante had died at the age of five, then bodies would not obey the first law of dynamics.

It is natural to say that (9) is *true*, and that (10) is *false*. Suppose A is "Dante died at five", that B is "We don't have the Divine Comedy", and C is "Bodies don't obey the first law of dynamics". Using symbolic notation, we would then say that while A ⋺→ B is true, A ⋺→ C is false. Now, A, B, C are all *false sentences*. If the counterfactual conditionals were truth-functional, (9) and (10) should have *the same* truth-value. This is so because in evaluating a sentence D whose main connective is true-functional, (a) *only* the truth-values of the sentences that constitute D count – plus, obviously, the truth table of the connective – and (b) the truth-values of these sentences uniquely determine the truth-value of D.

Since two counterfactuals can have different truth-values even if their antecedents and consequents have the same truth-values, it follows that the semantics of counterfactuals is not truth-functional.

7.3.3 The Ramsey Test for counterfactuals

In § 6.6 we have seen that it is possible to evaluate indicative conditionals on the basis of the so-called "Ramsey Test". The latter provides an alternative conception of indicative conditionals to that of the material conditional. With appropriate adjustments, such a test can be used to evaluate the truth-value of a counterfactual conditional. In this section, we present an example, then give a general definition, and discuss the differences with respect to the version for indicative conditionals introduced in § 6.6.

The basic idea of the Ramsey Test is that we accept a conditional as true on the basis of a "background" set of sentences. For convenience, we have conceived of this set as a "stock of information" or a "stock of beliefs". To determine whether we accept "If A, then B" as true, we must first add A to our information stock. In the version for counterfactuals we start with the same move, but in this case our information stock typically includes the *negation of antecedent* A, because counterfactuals are about scenarios contrary to the facts. Consequently, the simple addition of A would give us the contradiction A ∧ ¬A, and our information stock would become inconsistent. To avoid this, we need to add A, and make the necessary changes to the stock to keep it consistent. Then, we must determine whether B follows from the new consistent stock of information. If so, we accept counterfactual A ⋺→ B as true. Otherwise, it is false. An example will help us to understand more concretely how this works.

We wish to establish whether we have to accept as true that "L.H. Oswald did not kill John F. Kennedy ⋺→ No one else killed him", that is: "If L.H. Oswald had not killed John F. Kennedy, then no one else would have killed him". To do this, we must first identify the stock of information that correctly describes the piece of real world that interests us. Suppose that in the Kennedy case, things went as the Warren report relates. As a result, our information stock K will include:

(A) L.H. Oswald killed John F. Kennedy.
(B) L.H. Oswald was the only one to put John F. Kennedy's life at risk.
(C) There were no other plans to kill John F. Kennedy.

For convenience, suppose that K includes *only* these sentences, that is, in set notation: K = {A, B, C}. What happens now if we add ¬A ("L.H. Oswald didn't kill John F. Kennedy")? We get an *inconsistent* set of information, that is, a set that contains contradictory sentences. We must therefore *remove* A. We thus satisfy the consistency requirement we introduced above. We are now left with the set {¬A, B, C}. From these three sentences, it follows:

(D) Nobody killed John F. Kennedy.

We therefore accept that "If L.H. Oswald didn't kill John F. Kennedy ⊂→ No one else killed John F. Kennedy" is true.

We can define the Ramsey Test for counterfactuals, which we denote by RT*, as follows:

(RT*) To determine whether we must accept A ⊂→ B as true, add A to the information stock K, and (i) make the *minimum adjustments* necessary to maintain consistency; (ii) consider whether the new stock makes B true. If B is true, then we must accept A ⊂→ B as true. Otherwise, we must reject it as false.

We have already discussed the need for the new (counterfactual) stock to be consistent, but this definition adds something more: changes aimed at maintaining consistency must be, in some sense, *minimal*. Let us see why. Suppose that K = {A ∧ B, A, B}. Further, suppose we want to evaluate counterfactual ¬(A ∧ B) ⊂→ C. We add its antecedent to K. We need to remove the conjunction A ∧ B. Consequently, we have to remove *at least one* of A and B, because by keeping both of them we would obtain an *inconsistent* set: {¬(A ∧ B), A, B}. We therefore have three options: either we remove the set {A ∧ B, A}, or the set {A ∧ B, B}, or the set {A ∧ B, A, B}. In all these cases, the subsequent addition of ¬(A ∧ B) does not create inconsistencies. However, we tend to avoid the removal of {A ∧ B, A, B}. This is because the key to counterfactual reasoning is to change *just enough* with respect to the initial situation to maintain consistency and obtain the right set of sentences on the basis of which to establish whether the counterfactual is true or not.[4]

This characteristic of counterfactual reasoning seems to be due to an attitude of "cognitive economy", according to which we tend to minimally modify what is (or what we believe to be) true. To see that we actually have this attitude, let us re-

---

4. Whether we should remove A or B is another question. Suffice it to say here that there are no strictly logical criteria for this choice – unless, of course, either A or B is logically inconsistent with ¬(A ∧ B). In any case, the analysis of the criteria for this choice goes beyond the scope of this book.

turn to the Mr. Jones experiment for a moment. In constructing the counterfactual scenario, each of the participants modified, with respect to the story, only enough to obtain a scenario in which Mr. Jones did not die in the accident. No one, for example, proposed a scenario in which Mr. Jones took the usual route *and* the truck driver was not under the influence of drugs.

Note here the difference from the Ramsey Test for indicative conditionals. In the latter we also add antecedent A to K, but ideally we do so if neither A nor its negation are already included in our stock of information. In the case of RT*, instead, we add the antecedent A *despite* the initial stock containing ¬A. The Ramsey Test for counterfactual conditionals thus provides an intuitive idea of the criteria by which we accept (or do not accept) a counterfactual as true.

7.3.4 The semantics of counterfactuals

The semantics of counterfactuals is based on a rather intuitive idea, which we have actually already encountered in § 7.1: among the counterfactual scenarios, some are more similar than others to the real world. Consider, for example, the following scenarios s_1 and s_2:

> s_1: (i) there are no counterfeit products; (ii) only certain real-world buyers who purchase counterfeit Sennheiser headphones buy genuine Sennheiser products;
> s_2: (i) there are no counterfeit products; (iii) all buyers who purchase counterfeit Sennheiser headphones in the real world buy genuine Sennheiser products.

As we already said in § 7.1, scenario s_1 is *more similar to the real world* (let us call it s_0) than s_2. Since s_1 is more similar to the real world, in evaluating "There are no counterfeit products $\Box\!\!\rightarrow$ All those who have *actually* bought fake Sennheiser buy Sennheiser products" we must prioritize what is true in s_1 over what is true in s_2. Therefore, the conditional in question will turn out to be false, because in s_1, although the antecedent is true, the consequent is false.

These intuitive ideas receive rigorous treatment in the semantics of counterfactuals. In particular, to interpret counterfactual sentences, we take (i) a set S that includes the real world s_0 and an (even infinite) number of counterfactual scenarios $s_1, s_2, s_3, ...$; (ii) a comparative similarity relation with respect to each scenario s_i, which is interpreted as "x is at least as similar to s_i than y is";[5] (iii) a "relation of sat-

---

[5] We need to impose some formal properties on the similar relation in order for the semantics to work properly. The relation must be reflexive and transitive. Other properties can be imposed but we will not discuss them here. Notice that these properties are not essential for what we will say here on the counterfactual semantics.

isfaction" between sentences and scenario, which can be read as "A is true in s_i", or equivalently "in s_i it is true that A".[6] Then we define the following truth conditions for counterfactuals:

A ⊏→ B is true in s_i if and only if:
(1) there exists at least one scenario s_j in S where A is true and, for any scenario that is *at least* as similar to s_i as s_j, B is true if A is.[7]
Or if:
(2) there is no scenario s_j in S where A is true.

An equivalent formulation is:[8]

A ⊏→ B is true in s_i if and only if:
(1') given the set of scenarios S' in which A is true, B is true in all scenarios in S' which are more similar to s_i than any other scenario in S', or
(2') there is no scenario s_j in S where A is true.

Let us comment on these definitions before giving some examples. In both formulations, the second condition serves to handle counterfactuals with an antecedent whose truth is excluded from *any scenario in the model*. Consider, for example, counterfactual "2 + 2 = 4 and 2 + 2 = 5 ⊏→ Humanity has very few certainties". The second condition of both truth conditions makes this counterfactual vacuously true. While this sounds counterintuitive, consider that it is, in any case, different from what happens with material conditionals. In the case of the latter, a false antecedent is sufficient to make the conditional true; in the case of counterfactuals, this is not so, as we will see shortly. The insistence of logicians on this second condition is due to technical questions that do not interest us here.[9] However, since this condition is part of the classical semantics of counterfactuals, we have included it.

Regarding the first condition, it provides the results we had already intuitively illustrated by analyzing the example of Sennheiser products:

---

6. This relation is defined in such a way that any sentence containing negations, conjunctions, disjunctions, and material conditionals is truth-conditionally dependent on the truth-value of the sentences that compose it according to the truth table we saw in Chapter 5. Thus, the relation defines the property of "being true" in the possible scenarios.
7. By "at least as similar to s_i as s_j" we mean (i) "as similar as" or (ii) "more similar than". It is easy to see that if (i) is satisfied, B is also true in s_i.
8. Notice that both formulations of truth conditions presuppose that for every s_i in S, there is a subset S' of scenarios that are more similar to s_i than any other scenario external to S'. If this presupposition is false, truth conditions must be modified. In the following, we will take this presupposition for granted because it is intuitive and makes the semantics simpler. It goes beyond the educational purposes of this book to give a semantics that does not assume this presupposition.
9. In short, if we remove condition (2), both "A ⊏→ B" and "A ⊏→ ¬B" might be true in a scenario s_i. If this does not cause perplexity in the case of indicative conditionals (or at least in the case of material conditionals), it is counterintuitive in the case of counterfactuals.

> There are no counterfeit products $\mathbin{\Box\!\!\rightarrow}$ Only some of those who bought counterfeits in the real world buy genuine Sennheiser products.

Since there is at least one scenario where the antecedent is true and the scenarios where the antecedent and consequent are true are more similar to the real world than those where the antecedent is true and the consequent is false (in our case, we have defined a single scenario with these features), condition (1) is satisfied and therefore the counterfactual is true. Condition (1') basically says the same thing as condition (1).

Suppose that p is "There are no counterfeit products", and that q is "Only some of those who bought counterfeits in the real world buy Sennheiser products". Let us try to establish the truth of $p \mathbin{\Box\!\!\rightarrow} q$ in the real world s_0. Suppose that $S = \{s_0, s_1, s_2\}$ is the set of possible scenarios, that s_1 is more similar to s_0 than s_2, and that:[10]

- in s_0 it is true that: $\neg p, \neg q$;
- in s_1 it is true that: p, q;
- in s_2 it is true that: $p, \neg q$.

This model formally captures the considerations we made when discussing the Sennheiser case. According to this model, $p \mathbin{\Box\!\!\rightarrow} q$ is true in s_0. Suppose now that r is "*Everyone* who bought counterfeit Sennheiser products buys real Sennheiser products". We want to establish whether $p \mathbin{\Box\!\!\rightarrow} r$ in s_0 (the real world). For convenience, we will do so by building a model different from the previous one, but which – like the previous one – has $S = \{s_0, s_1, s_2\}$. However, s_2 happens to be more similar to s_0 than s_1, and:

- in s_0 it is true that: $\neg p, \neg r$;
- in s_1 it is true that: p, r;
- in s_2 it is true that: $p, \neg r$.

It follows that $p \mathbin{\Box\!\!\rightarrow} r$ is false in s_0, because for *every* scenario where p is true and r *is true* (i.e. s_1) there is a scenario *more similar* to s_0 where p is true and r *is false* (this scenario, of course, is s_2).

This last example also helps us to discuss a further issue. In the model we have just presented, the similarity ordering we have imposed (s_2 is closer to s_0 than s_1) has an obvious intuitive grip: a scenario where p and $\neg r$ are true is more similar to the real one than a scenario in which p and r are true, as we said in § 7.1. However, not every model necessarily presents similarity relations that have this intuitive grip. Let us see why.

Given a set S of scenarios, it is possible to define on it a large number of different relations of comparative similarity. In the last model we discussed, s_2 is closer to

---

10. We can refrain from specifying whether s_0 or s_1 is more similar to s_2, and whether s_0 or s_2 is more similar to s_1, because this is not relevant for establishing the truth of the counterfactual in s_0.

7 Conditional reasoning, II: The counterfactual conditional

s_0 than s_1. But we could define a different model in which we have the same scenarios and the same true sentences, but a different order. In this alternative model, s_1 is more similar to s_0 than to s_2. This is because the similarity relations are, in principle, independent of which sentences are true in the scenarios considered. Consequently, we can have models in which the similarity ordering with respect to a given scenario has an intuitive grip – because the set of the true sentences in the various scenarios makes that ordering intuitive, as in the last example we discussed – and, conversely, we can have models in which the similarity ordering is not natural – because the set of true sentences in the various scenarios makes it counterintuitive.

This feature is actually an advantage for the semantics of counterfactuals because it provides an explanation for our disagreement with the same counterfactuals. This disagreement is usually sharper than that regarding indicative conditionals. Consider this example:

> If the *New York Times* had discovered the relationship between Trump and Putin before the 2016 presidential election, then Trump would not have won the 2016 presidential election.

We look for a model where this counterfactual is true. In particular, suppose that p stands for the antecedent of this counterfactual, and q for the consequent. We find that:

- in s_0 it is true that: $\neg p, \neg q$.

Since we have only two sentences, we have four possible combinations of their truth-values. Each needs a scenario to be represented. For example, s_0 represents the real world, where p and q are both *false*. Note, however, that we are only interested in scenarios where p is true. In short, we only need *two* scenarios beyond s_0:

- in s_1 it is true that: $p, \neg q$;
- in s_2 it is true that: p, q.

We have three possible options regarding the similarities of these two scenarios with respect to s_0. Either "s_2 is more similar to s_0 than s_1", or "s_1 is more similar to s_0 than s_2", or "s_1 and s_2 are equally similar to s_0".

In the model where s_2 is more similar to s_0 than s_1, $p \,\Box\!\!\rightarrow q$ is true in s_0. In this situation, a scenario in which there is a huge scandal about a political candidate and that candidate loses the election is more similar to the real world than one in which there is a scandal and the candidate wins the election.

In the model where s_1 is more similar to s_0 than s_2, $p \,\Box\!\!\rightarrow q$ is *false* in s_0, while $p \,\Box\!\!\rightarrow \neg q$ is true in s_0. In this situation, the result of the presidential election does not change even if a scandal breaks out. Or, more precisely: a scenario in which a scandal breaks out and Trump is elected anyway is more similar to reality than the scenario in which the scandal breaks out and Trump loses the election. This is a "pessimistic" point of view, but it is something *logically possible*.

In the model in which the two scenarios are equally similar to s_0, both $p \mathrel{\square\!\!\rightarrow} q$ and $p \mathrel{\square\!\!\rightarrow} \neg q$ are false in s_0. This model is neutral on which is most similar to the real world.

A point to emphasize is the following: all these arrangements are possible, and each of them seems to incorporate a vision of "what is most similar to reality". The first ordering is, so to speak, "optimistic", the second "pessimistic", and the third "neutral". It is easy to see that these three orderings reflect three different approaches that we could intuitively adopt to evaluate the counterfactual $p \mathrel{\square\!\!\rightarrow} q$. In other words, an agent might hold the counterfactual to be true because he believes that Trump's defeat in the event of a scandal *would be* more in line with what usually happens in these cases; another might believe it to be false because the civic spirit of US voters has now vanished; still another might be neutral about whether Trump would win or lose if a scandal broke up. Obviously, the disagreement between these three points of view *is not a disagreement* on the *facts* – everyone knows that p and q are false in reality.

Counterfactual semantics helps us understand *on what* the disagreement is about: not about the real world but about *what is most similar to the real world*.[11] Since there are many views on it, there is also a deep disagreement on several counterfactuals. Since we can define many similarity relations on the same scenarios, this is a strong point in favor of the semantics of counterfactuals. The latter reflects the plurality of conceptions of "what is most similar to the real world" and explains our disagreements about counterfactuals.

Let us now see the connection between this way of dealing with counterfactual semantics and the "intuitive" method of accepting or rejecting a counterfactual through the Ramsey Test (RT*). Take A $\mathrel{\square\!\!\rightarrow}$ B. The counterfactual scenarios in which A is true have a function similar to the stocks of information to which we "add" A. These scenarios are consistent, and therefore in them neither ¬A nor anything that implies ¬A is true. This reflects the consistency condition we want our information stocks to fulfil. Moreover, the semantics of counterfactuals incorporates a sort of "economy" condition. The condition of "maximum possible similarity" to the real world corresponds to the condition of "minimum adjustments necessary to maintain consistency": it suggests that we change only what is minimally necessary to evaluate the counterfactual.

There is, however, a crucial difference between the two conditions. Let us see why. We shall start from the Ramsey Test (RT*) and the condition that we must remove *exclusively* what would contradict the information A just added. This means leaving everything that was included in the initial stock and that is compatible with A. Now, let us take "*NYT* discovers relations between Putin

---

11. Thus, indirectly, the disagreement concerns the *model* by means of which counterfactuals are evaluated.

7 Conditional reasoning, II: The counterfactual conditional

and Trump before the 2016 US presidential election $\mathrel{\Box\!\!\to}$ Trump does not win the 2016 US presidential election", and apply the Ramsey Test. Suppose that our initial stock of information is $\{\neg p, \neg q\}$. We add p (the antecedent of the counterfactual) to this stock and remove $\neg p$. We have made the minimum changes necessary to ensure consistency, so we do not have to do anything else. We remain with $\neg q$: Trump wins the 2016 US presidential election. The Ramsey Test returns $\{p, \neg q\}$ as the only possible result. This means that, according to the test, $p \mathrel{\Box\!\!\to} q$ is not acceptable as true.

However, the semantics of counterfactuals permits the building of at least one model in which $p \mathrel{\Box\!\!\to} q$ is true (with respect to s_0), and that this model seems to us something possible, not something to be ruled out as meaningless, abnormal or unreasonable.

What explains this divergence? The semantics of counterfactuals tolerates changes of *other* kinds than those to maintain consistency, as long as they are *in some sense* minimal. To appreciate this, let us take the model on the basis of which we established that $p \mathrel{\Box\!\!\to} q$ is true in s_0 (where this counterfactual is: "If the *NYT* had discovered relationships ..., then Trump would not have won the 2016 presidential election"). In that model, s_2 is the scenario where the *NYT* discovers the relationships and Trump does not win. This scenario is more similar to s_0 than s_1. Note, however, that it is s_1 that makes the minimum changes necessary to maintain consistency: indeed, the fact that Trump wins is compatible with the fact that the *NYT* discovers his relations with Putin. Conversely, s_2 makes *more* changes than s_1, since in s_2 q (instead of $\neg q$) is true. Most importantly, the latter change is not *necessary* to maintain consistency.

At the same time, however, only relevant changes are permitted. A scenario in which the *NYT* discovers the relations between Putin and Trump, Putin exercises telepathic powers to persuade the entire American electorate and Trump wins, is a scenario that is *too dissimilar* to the real one to take it into account. In other words, similarity relations help us to ensure that some scenarios cannot really matter in the evaluation of counterfactuals, because they are too dissimilar to the real scenario. When we build models for a given example, these scenarios will typically be ones where a *lot of things* are different from the real facts. In short, being less precise than the Ramsey Test (RT*), the semantics of counterfactuals follows the intuition that, in order to be truly relevant to the evaluation of a counterfactual, a scenario must make as *many* changes *as necessary* to address the problem. The point is that, contrary to the Ramsey Test (RT*), in the semantics of counterfactuals there is no method to establish *a priori* what "enough" means: it must be examined case by case.[12]

---

12. Notice that the semantics of counterfactuals seem to reflect the changes we tend to make: we change exceptional events, our actions, the events we can control, etc. (see § 7.2).

7.3.5 Fallacies of counterfactual reasoning

Let us now consider some fallacies of counterfactual reasoning. These fallacies are particularly interesting because they follow inference schemes that are valid for material conditionals. Such schemes are known as the *truth of the consequent, the strengthening of the antecedent, contraposition, transitivity*. The fact that they are valid for material conditionals but not for counterfactuals helps us to perceive the differences between the two kinds of conditionals more clearly. This also partly explains why we tend to follow these schemes when we reason with counterfactuals, even if they are not valid. To see that such schemes do not hold for counterfactuals, we need to construct counterexamples. In doing so, we will put into practice what we have observed so far regarding the semantics of counterfactuals. Remember that in the notation used here, A → B indicates the material conditional.

The truth of the counterfactual does not follow from the truth of the consequent. As we saw in Chapter 6, the truth of the consequent of a material conditional is sufficient to establish the truth of the conditional itself. In other words, the following inference rule is valid:

$$\frac{B}{A \to B}$$

This principle does not seem plausible for counterfactual conditionals. After all, the fact that a sentence is true does not guarantee that it *would have been* true if things had turned out differently. Consider, for example, the following sentence:

> Trump won the US presidential election in 2016.

This sentence is true. But it does not follow from this that:

> If the *New York Times* had discovered relations between Putin and Trump before the election, Trump would have won the presidential election in 2016.

According to the semantics of counterfactuals, inferences of the following form *are not deductively valid*:

$$\frac{B}{A \mathbin{\Box\!\!\to} B}$$

Let us give a counterexample. To do this, we must show that there is a true sentence B for which A ⟼ B is not true. Suppose that p is "Trump dies before the 2016 elections" and q is "Trump wins the US presidential election in 2016". In the real world (scenario s_0), q is true and p is false. That is:

- in s_0, it is true that: $\neg p, q$.

To evaluate the counterfactual, we must consider only scenarios in which p is true. This leaves us with two possible scenarios: s_1, where p and q are both true, and s_2, where p is true and q is false. In short, in s_1, Trump dies before the elections and Trump wins anyway; in s_2, on the other hand, Trump dies before the elections and Trump does not win.

Is s_1 or s_2 more similar to the real world? Given the ways in which candidates are chosen, it is reasonable to consider s_2 more similar to s_0 than s_1. In fact, it seems very reasonable to think that, following the death of Trump, the Republican Party would have nominated another politician for the US presidency rather than insisting on nominating a deceased person. Consequently, $p \mathrel{\Box\!\!\rightarrow} \neg q$ is true, and, therefore, $p \mathrel{\Box\!\!\rightarrow} q$ is false. Since q is true, this is enough to construct a case in which the above inference is not valid (since q is true, but $p \mathrel{\Box\!\!\rightarrow} q$ is not).

The strengthening of the antecedent is not valid. For indicative conditionals, a principle known as "strengthening of the antecedent" applies. According to this principle, if "A → B" is true, we can strengthen the antecedent with any sentence C, obtaining another true conditional:

$$\frac{A \rightarrow B}{(A \wedge C) \rightarrow B}$$

A glance at the truth table of the conditional is enough to see that this rule of inference is deductively valid. Consider, for instance, this indicative conditional:

(11) If a triangle is isosceles, then it has at least two congruent angles.

We know that (11) is true. However, if (11) is true, also is:

(12) If a triangle is isosceles *and is blue*, then it has at least two congruent angles.

The same rule of inference, however, is not valid in the semantics of the counterfactuals we have introduced. Let us assume that (13) is true.

(13) If the *NYT* had discovered some relationships between Trump and Putin before the 2016 US presidential election, then Trump would *not* have won the 2016 US presidential election.

And now consider the following counterfactual, obtained by "strengthening" the antecedent of (13):

(14) If the *NYT* had discovered the relationships between Trump and Putin before the 2016 US presidential election *and kept them secret*, then Trump would not have won the 2016 US presidential election.

It is reasonable to consider this conditional *false*. If the relationships between Trump and Putin had been kept secret, the electorate would not have changed its opinion, and therefore we have no reason to think that the 2016 presidential election would have had a different outcome. These intuitive considerations help us to see that the following inference is not valid:

$$\frac{A \mathrel{\Box\!\!\rightarrow} B}{(A \land C) \mathrel{\Box\!\!\rightarrow} B}$$

We can build a counterexample to it. Again, suppose p is "The *NYT* finds out about Trump-Putin relationships before the 2016 US presidential elections", but also assume that q is now "Trump does not win the 2016 US presidential election". Suppose further that r is "The *NYT* keeps Trump-Putin relationships secret".

Again, in s_0 (the real world) p, q, and r are all false – the *NYT* did not discover the relationships in question before the election and it did not keep them secret; besides, Trump won the 2016 presidential election. To construct a counterexample to the inference scheme introduced above, we must start from a model in which $p \mathrel{\Box\!\!\rightarrow} q$ is true, but $(p \land r) \mathrel{\Box\!\!\rightarrow} q$ *is* not true. We need four counterfactual scenarios: two in which p is true and r *is false* (in one q is true, in the other it is false), and two in which the conjunction $p \land r$ is true (in one r is true, in the other it is false). We define the four scenarios as follows:

- in s_1 it is true that: $p, \neg r, \neg q$;
- in s_2 it is true that: $p, \neg r, q$;
- in s_3 it is true that: $p \land r, q$;
- in s_4 it is true that: $p \land r, \neg q$.

Those scenarios in which p is true but r is *not* (i.e. s_1 and s_2) are more similar to s_0 than the two scenarios in which p and r are both true (i.e. s_3 and s_4). After all, a situation in which an important newspaper gets a scoop and divulges it seems more similar to how things go in reality, compared to a scenario in which the same newspaper gets a scoop but does not exploit it in any way. Finally, s_2 seems more similar to s_0 than s_1 – at least if we consider how the electorate tends to react to certain kinds of news items. Because of this arrangement among the four scenarios, we must conclude that $p \mathrel{\Box\!\!\rightarrow} q$ is true.

Let us now consider $(p \land r) \mathrel{\Box\!\!\rightarrow} q$. If we restrict our attention to the scenarios in which the antecedent is true (s_3 and s_4), the scenario in which Trump wins the election (i.e. s_4) looks more similar to the real world s_0 than the scenario in which Trump does not win (i.e. s_3): since the news about his relations with Putin have not been spread, the opinions of the electorate are in s_4 exactly as they are in the real world s_0. Because of this, $(p \land r) \mathrel{\Box\!\!\rightarrow} q$ is *false*. This is sufficient to produce a counterexample to the inference scheme we are discussing.

Counterfactuals do not support contraposition. Another scheme valid in the case of indicative conditionals but invalid in the case of counterfactuals is *contraposition*. In this scheme, the antecedent and the consequent of a material conditional are inverted and negated. We might easily check that this scheme is valid on the basis of the *modus tollens* and the rule of introduction of the conditional:

$$\frac{A \to B}{\neg B \to \neg A}$$

We can also check the validity of the scheme using the truth table of material conditionals.[13] Here is an example. From sentence (11) above, it follows:

(15) If a triangle does not have at least two congruent angles, then it is not isosceles.

This kind of inference is not valid for counterfactual conditionals. Again, let us take an example and discuss what we would intuitively say about the conditionals involved. Then, we will show that the semantics of counterfactuals conforms to our intuitions in this regard. Suppose that Ann wants to follow Claude wherever he goes, because she has a crush on him. Claude has been invited to a party, and he likes parties very much. Ann heard about it and decides to go to the party, thinking that Claude will go there. However, Claude feels more at ease when he does not see Ann and therefore, he has decided not to go to the party. Given this situation, we would accept the following counterfactual as true:

(16) If Claude had gone to the party, Ann would have gone.

But we would reject the following as false:

(17) If Ann hadn't gone to the party, Claude wouldn't have gone.

Let us now construct a model showing the invalidity of the following inference scheme:

$$\frac{A \mathbin{\Box\!\!\to} B}{\neg B \mathbin{\Box\!\!\to} \neg A}$$

Suppose p is "Claude goes to the party" and q "Ann goes to the party". In the real world (s_0), p is false and q is true. We now need three counterfactual scenarios. In

---

13. Suppose that B is true. In this case, A → B is true because the consequent is true. Further, since ¬B is false, ¬B → ¬A is true because it has a false antecedent. Suppose now that A is false. In this case, A → B is true because it has a false antecedent. Further, since ¬A is true, ¬B → ¬A is true because it has a true consequent. Since the cases in which B is true and those in which A is false exhaust the cases in which A → B is true, this reasoning suffices to prove the validity of the above inference scheme.

the first two p is true (in one of them q is true, in the other q is false); in the third both p and q are false:

- in s_1 it is true that: p, q;
- in s_2 it is true that: $\neg q, p$;
- in s_3 it is true that: $\neg q, \neg p$.

Now, s_1 and s_2 are scenarios in which Claude goes to the party. Scenario s_1, in which Ann also goes to the party, is more similar to the real world than s_2, in which Ann does not go to the party (in fact, in the real world Ann *goes* to the party). This is enough to conclude that $p \mathrel{\Box\!\!\rightarrow} q$ is true. Now, consider the two scenarios in which Ann does not go to the party: s_2 and s_3. In the first one, Claude goes to the party, in the second one he does not go. Given what we have said about Claude's preferences, it is reasonable to assume that scenario s_2 is more similar to the real world than s_3. Hence, $\neg q \mathrel{\Box\!\!\rightarrow} \neg p$ is false. This is enough to show that the inference scheme we are discussing is invalid.

Counterfactual conditionals do not support transitivity. The following inference scheme:

$$\frac{\begin{array}{c} A \rightarrow B \\ B \rightarrow C \end{array}}{A \rightarrow C}$$

is valid for the material conditional. We can thus say that the material connective has the property of transitivity. Here is an example. From the following two conditionals:

(18) If a triangle is isosceles, then it has at least two congruent angles;
(19) If two angles are congruent, then they have the same width,

the following conditional clearly follows:

(20) If a triangle is isosceles, then it has at least two angles with the same width.

Counterfactual conditionals, however, are not transitive. Let us take Claude, Ann, and Frances. Claude and Ann had an affair, but now Claude is in love with Frances. Ann wants to try to win Claude back, but she does not want to run the risk of meeting Frances. Now consider these counterfactuals:

(21) If Frances had gone to the party, Claude would have gone too;
(22) If Claude had gone to the party, Ann would have gone too.

We are led to believe that both counterfactuals are true. The first one because Claude wants to have the opportunity to meet Frances. The second one because

Ann wants to have the opportunity to win back Claude. But we would say that the following conditional:

(23) If Frances had gone to the party, Ann would have gone too

is false. As we have said, Ann does not want to meet Frances. This example shows that counterfactual reasoning cannot be transitive.

Let us construct a counterexample for the following inference scheme:

$$\frac{A \mathrel{\Box\!\!\rightarrow} B \\ B \mathrel{\Box\!\!\rightarrow} C}{A \mathrel{\Box\!\!\rightarrow} C}$$

This counterexample will be a little more complex than the previous ones. Suppose p is "Frances goes to the party", q is "Claude goes to the party", r is "Ann goes to the party". In s_0, the real world, p, q, r are all false. We have three sentences (p, q, r), each of which can be either true or false. There are therefore $2^3 = 8$ combinations of truth-values. One of these combinations coincides with the real world s_0, where $\neg p$, $\neg q$, $\neg r$ are true. The other seven are satisfied by the following scenarios:

- in s_1 it is true that: $p, q, \neg r$;
- in s_2 it is true that: p, q, r;
- in s_3 it is true that: $p, \neg q, \neg r$;
- in s_4 it is true that: $p, \neg q, r$;
- in s_5 it is true that: $\neg p, q, r$;
- in s_6 it is true that: $\neg p, \neg q, r$;
- in s_7 it is true that: $\neg p, q, \neg r$.

Suppose, in addition, that Frances does *not* like going to parties, and that Frances is indifferent both to Claude's feelings and to Ann's dislike for her. Let us seek an ordering of this situation that makes both $p \mathrel{\Box\!\!\rightarrow} q$ and $q \mathrel{\Box\!\!\rightarrow} r$ true. Let us consider the four scenarios in which Frances goes to the party (p). They are s_1-s_4. We postulate that the closest to the real world s_0 is the one in which Claude goes to the party but Ann does not – Ann's priority, we suppose, is not to meet Frances, which is more important to her than the opportunity to meet Claude. Hence, s_1 is more similar to s_0 than s_2, s_3, s_4. This makes $p \mathrel{\Box\!\!\rightarrow} q$ and $p \mathrel{\Box\!\!\rightarrow} \neg r$ true. Now let us take the four scenarios in which Claude goes to the party (q) – i.e. s_1, s_2, s_5, s_7. We must complete the arrangement so that (i) $q \mathrel{\Box\!\!\rightarrow} r$ is true, (ii) the ordering remains consistent with the truth of $p \mathrel{\Box\!\!\rightarrow} q$. Among the scenarios where q is true, s_5 is more similar to s_0 than s_7. What about the other ones? Well, s_7 is more similar to s_0 than both s_1 and s_2: in s_7 Frances does not go to the party, and this is more in agreement with the fact that she does not like parties. Since this is independent of what Ann and Claude do (Frances is indifferent to both), we can reasonably assume s_5 is more similar to s_0 than *any* scenario where "Frances goes to the party" (p) is true, including both s_1 and s_2. The fact that s_5 is more similar to s_0 than the other scenarios where q is true guarantees the truth of $q \mathrel{\Box\!\!\rightarrow} r$. Finally, we have seen that s_1 is more similar to s_0 than all other scenarios where p is true. In s_1, however, r is false. Consequently, there is no scenario where (i) p is true, (ii) r is true, and (iii) which is more similar to s_0 than the scenarios where p is true but r is false. Indeed, s_2 and s_4 are the only ones satisfying (i) and (ii), and they are both *less* similar to s_0 than s_1, which satisfies (i) but not (ii). We have then built a model in which $p \mathrel{\Box\!\!\rightarrow} q$ and $q \mathrel{\Box\!\!\rightarrow} r$ are true, and yet $p \mathrel{\Box\!\!\rightarrow} r$ is false. This is sufficient to show that transitivity is not a property of counterfactuals.

Notice that counterexamples to an inference scheme do not prove that in *all* models where the premises are true, the conclusion is false. They simply show that *there is at least one* model in which the premises are true and the conclusion false. This, in any case, is sufficient to show that there is no guarantee of passing from true premises to a true conclusion using it; this guarantee is however expected from deductive reasoning.

7.4 Counterfactuals and causality

In § 7.2 we mentioned the fact that we often consider counterfactual alternatives when we must evaluate the presence (or absence) of causal links. Here we will briefly discuss the relationships between counterfactual and causal reasonings. Imagine that you want to establish whether there is a causal relationship between two events E_1 and E_2, that is, whether "E_1 causes E_2" or "E_2 causes E_1" is true. To solve this problem, we often ask: "If E_1 had not happened, would E_2 have happened?" If the answer to this question is yes, we are inclined to conclude that E_1 *is not* the cause of E_2. If the answer is negative, we would be inclined to conclude that E_1 *is* the cause of E_2. We ask a similar question if we want to establish whether E_2 causes E_1. This shows that we often use counterfactual reasoning to ascertain causal connections.

Before considering this topic in more detail, it is worth making some clarifications. First, a counterfactual involves *sentences*; in the example above, in particular, counterfactuals involve sentences that refer to events (E_1 and E_2). However, causal relationships do not involve sentences; rather they involve *events* or *facts*. This leads to significant differences. We say, "If a stone hit a window, then it would break", but not "The stone hits the window causes the window to break". Instead, we say "The impact of a stone causes a window to break". Using a more technical turn of phrase (somewhat different from our everyday language), we could say "The event that the stone hits the window causes the event that the window breaks". We have created two nominal constructions ("The event that the stone hits the window" and "The event that the window breaks") starting from two sentences ("The stone hits the window" and "The windows breaks"). This type of construction allows us to discuss the relationships between a possible causal link between events, and the counterfactual links between the sentences that refer to the events in question. Here we will use the schematic letters $E_1, E_2, E_3, \ldots$ for sentences expressing events, and the symbols $!E_1!, !E_2!, !E_3!, \ldots$ for the nominal constructions obtained from $E_1, E_2, E_3, \ldots$ So, if E_1 is the sentence "The stone hits the window", $!E_1!$ will be the nominal construction "The event that the stone hits the window". $!E_1!$, unlike E_1, can legitimately be the predicate term in causal links. Second, we will use X as a variable for sentences. In other words, X can take sentences as values. For example, in "If X, then Julius will not pass the exam", values for X could be "Julius does not study", or even "the Government falls", or "cold nuclear fusion is very difficult to obtain". All these *can* be values of X because they are sentences. Obviously, not all of them make "If X, then Giulio will not pass the exam" true. We must select those values that make this sentence true. Given X, we use !X! to denote the nominal construction obtained from X. The value of X determines the value of !X!.

7.4.1 The counterfactual test

We have said that there is a relationship between causal and counterfactual links. But what exactly does this relationship consist of? It is not easy to tell. Here we will discuss an apparently intuitive view, which however does not turn out to be completely correct. This position identifies counterfactuals and causal links between events. The *counterfactual Test* (CT) expresses this view:

(CT) $!E_1!$ causes $!E_2!$ *if and only if* $\neg E_1 \mathbin{\square\!\!\rightarrow} \neg E_2$

For convenience, we can say that $!E_2!$ *counterfactually depends* on $!E_1!$ if and only if it is true that: $\neg E_1 \mathbin{\square\!\!\rightarrow} \neg E_2$. (CT) therefore says that $!E_1!$ causes $!E_2!$ if and only if $!E_2!$ counterfactually depends on $!E_1!$. Note that the truth of

$$\neg E_1 \mathbin{\square\!\!\rightarrow} \neg E_2$$

is independent of the truth of

$$\neg E_2 \mathbin{\square\!\!\rightarrow} \neg E_1$$

since the first can be true and the second false. For example, "If Claude hadn't gone to the party, Ann wouldn't have gone" may be true, but "If Ann hadn't gone to the party, Claude wouldn't have gone" may be false. So, according to (CT), $!E_1!$ can cause $!E_2!$ without $!E_2!$ causing $!E_1!$. This guarantees (CT) to be compatible with the fact that causal relationships are not symmetrical. This is important for the adequacy of (CT). In the next section, we will see, however, that this is not enough, and that (CT) is not a completely adequate principle.

7.4.2 Inadequacy of the counterfactual test

(CT) is conceptually inadequate, i.e. it does not correctly describe the relationship between counterfactuality and causality. A simple example shows us that counterfactual dependence is not a necessary condition for causality, that is, we can have causal links without counterfactual dependence, in contrast to what (CT) states. The fact that counterfactual dependence is not necessary for causation means that *there can be a causal relationship between A and B even without there being a counterfactual dependence between A and B*. Consider the following example:

> *Bill and Suzy.* Bill and Suzy both want to throw a stone at a glass bottle. Suzy throws first or harder. Her stone comes first. When Bill's stone reaches where the bottle was, only glass shards are there. However, Bill's stone had enough strength to break a bottle.

In this example, Suzy's throw is clearly the cause of the breaking of the bottle, but there is no counterfactual dependence between the two events. As the story goes, the following counterfactual is false:

> If Suzy hadn't thrown the stone, the bottle would not have broken.

This is because if Suzy had not thrown her stone, Bill's stone would have hit the bottle and broken it. We are therefore in a situation where !E_1! (Suzy's throw), !E_2! (the breaking of the bottle) and !E_3! (Bill's throw) are such that

$$!E_1! \text{ causes } !E_2!$$

but the following counterfactual is *false*:

> If E_1 had not happened, E_2 would not have happened.

In other words, it is false that:

> E_2 is counterfactually dependent on E_1.

Counterfactual dependence is therefore not necessary for causality. However, in the simplest and most common cases, (CT) works because there are usually no potential competing causes, as in the example of Bill and Suzy. Where these kinds of situations arise, however, (CT) does not prove to be a good principle. Furthermore, we must not take it for granted that these kinds of situations are of little importance from a practical point of view.

7.5 Conclusion

In this chapter, we have discussed counterfactual conditionals and the forms of deductive reasoning associated with them. We have explained what counterfactuals are, and what kind of cognitive capacity we employ to grasp their meaning. We have also reviewed some experimental results on reasoning with counterfactuals, specifying the importance of the latter in our daily activities and the importance of these descriptive results for our normative approach. In addition, we have dealt with the *logic* of counterfactuals, returning to a normative point of view, and focusing on their differences from indicative conditionals. Moreover, we have discussed their truth conditions and their semantics. We have closed this chapter by discussing the relationship between counterfactual and causal reasoning.

Part III
Non-deductive arguments

8 | Reasoning with explanatory hypotheses

From this chapter on we will analyze some kinds of non-deductive reasoning. As we said before, in Chapter 2, an argument is non-deductive when the truth of the premises does not guarantee the truth of the conclusion but only its probability or plausibility. In this chapter, we will analyze the forms of non-deductive reasoning used to explain or identify the causes of an event or fact. That this event has occurred is stated in the premises of the reasoning, and its conclusion consists in the explanation or cause of that event. The kinds of reasoning that we will analyze below seek to establish, among the various possible explanations for or causes of an event, which of them is the most probable. We will start with a case in the history of medicine (§ 8.2); we will then discuss some basic aspects of *abductive reasoning* and eventually focus on the *inference to the best explanation*, a form of reasoning that seeks to identify the best explanation for a fact or event out of those available (§ 8.3). Finally, we will look at some specific applications of inference to the best explanation by discussing *Mill's methods* for causal reasoning (§ 8.4). Indeed, these methods provide procedures for determining which is the best of a set of hypotheses regarding what causes what.

8.1 Introduction

Both in our daily experience and in scientific research we often need to explain facts whose causes or reasons are not known. For example, I may wonder why my house keys are not in the trinket tray where I usually leave them, and virologists and epidemiologists may want to understand why a given disease seems to have a higher mortality rate in certain geographic areas than others. Some economists may wonder why a sub-prime crisis broke out between 2006 and 2008. And so on. To explain these facts, we must develop *explanatory hypotheses*, that is, the possible explanations for a fact or event. For example, I can assume that I left the keys in my jeans for once; or that in different geographical areas the statistical survey of deaths from a disease is carried out in different ways, with consequences on the determination of the mortality rate of the disease; or that investors had not ascertained the solvency of the loans before taking a gamble on securities that incorporated sub-prime mortgages.

There are many logical models that show us how an explanation should be "represented". Typically, an explanation aims to identify explanatory links between a sentence we call *explanandum*, which expresses what we have to explain, and one or more sentences called *explanans*, which instead express what would explain the *explanandum*. Among the many available, there is also a *deductive* model of explanation, according to which an explanation is a particular type of argument in which the *explanandum* necessarily follows from the *explanans*; that is, for the deductive model of explanation, the latter is a particular type of deductive argument. However, in formulating hypotheses, this path usually does not interest us: we already know that the *explanandum* is true, and what we are looking for is an *explanans* that gives us reasons, even non-conclusive ones, for the *explanans*. That is, we proceed from the *explanandum* to the *explanans*, assuming the former as a premise (or as one of the premises) and the latter as a conclusion. This makes our reasoning, in fact, non-deductive. A reasoning that seeks to identify an *explanans* of a certain fact or event is called *abductive*. Starting from our *explanandum* we can go back to several possible explanations which provide reasons for inferring the truth of the *explanandum*. The next step will consist in establishing which of the various possible explanations is the best one. But to understand all this, let us start with an example.

8.2 A medical case

Starting out from a true story in medicine, let us try to reconstruct the two fundamental steps in explanatory reasoning: the formulation of hypotheses, and the search for the hypothesis that best explains the *explanandum*. Between 1846 and 1847 a Hungarian doctor, Ignác Semmelweis, conducted an investigation to discover the origin of puerperal fever, a serious infection of the uterus that can occur after childbirth and can have fatal effects. At the time of Semmelweis' investigations, the existence of contamination by pathogens such as bacteria was not yet known; this is what actually causes the disease. There were various hypotheses regarding the causes of puerperal fever at the time. Here are some of them:

> *Hypotheses*
> H1: Puerperal fever is caused by "epidemic influences" that periodically occur in entire districts of a country or city.
> H2: Puerperal fever is contracted in overcrowded places.
> H3: The contraction of puerperal fever depends on the diet.
> H4: Puerperal fever depends on the "lochia", that is from secretions that come out of the vagina for a period of about 6 weeks after delivery, which are not expelled and have putrefied as a result of stagnation, which rise up in the tissues and blood, causing pain, fever and finally the death of the woman.
> H5: During the nine months of pregnancy, the veins could absorb poisons produced by accumulated fecal matter. The cause of the accumula-

tion was attributed to the uterus which, when enlarged, exerts pressure on the intestine, causing a stasis.

Semmelweis was struck by the fact that in the Vienna General Hospital, where he worked, puerperal fever affected women admitted to the First Maternity Ward (FMW) far more widely than women admitted to the Second Maternity Ward (SMW). The following was the case:

Problem
In the Vienna General Hospital, many more women in FMW contract puerperal fever than in the adjacent SMW. *Why? What* explains *the difference in the mortality rates of the two wards?*

Semmelweis tried to analyze the conditions of the two wards and obtained the following:

Initial evidence
E1: There is a significant difference between the mortality rates in the First Maternity Ward and the Second Maternity Ward.
E2: The crowding rate of the two wards is the same.
E3: The diet in the two wards is the same.

This evidence disproves hypotheses H1-H5. More precisely, the fact that the two wards were adjacent denies H1: if the disease were caused by epidemic influences that periodically affect cities or entire regions, it should have affected both wards *equally, or in any case in very similar proportions*. H1, therefore, does not explain the difference in mortality rates. E2 also denies H2, while E3 denies H3, for reasons similar to those we have just seen. The same is true for H4 and H5: if the fever depended on the "lochia" or on the matter accumulated in the veins, it should have affected both wards indifferently. Therefore, these hypotheses, too, were unable to explain the fact investigated by Semmelweis, namely the difference in the mortality rates. They failed to explain it because, given the initial evidence E1-E3, they instead led to the opposite conclusion, namely a *uniformity* in mortality rates, which in fact did not occur. The evidence E1-E3 collected by Semmelweis was therefore useful to *disprove*, or at least to make *less plausible*, some explanatory hypotheses. It is also clear that E1-E3 *did not follow* from those hypotheses, in the sense that E1-E3 were not *deducible* from them. The hospital managers then formulated a new hypothesis that could explain the onset of puerperal fever, a hypothesis not disproved or weakened by E1-E3:

H6: Fever is caused by medical students, who receive their obstetrical training in the FMW but not in the SMW. Due to inexperience, students are probably rougher when they examine patients, and this can lead to injuries and the contraction of puerperal fever. The midwives who work in SMW, on the other hand, use better procedures and do not expose themselves to the same risk of injuring patients.

However, in the course of his investigation, Semmelweis gathered further evidence that allowed him to judge H6 implausible:

> E4: Medical students and midwives use exactly the same methods to examine patients in both wards.
> E5: A hospital doctor, once injured with a scalpel he was using in the autopsy room, developed the same symptoms seen in puerperal fever victims.

Now, E4 disproves H6. E5 instead suggests the formulation of a new hypothesis, exactly the one formulated by Semmelweis to explain the origin of puerperal fever and all the evidence collected by him:

> H7: Puerperal fever is produced by "infectious materials" (today we would say "bacteria") that are introduced into the wounds produced by childbirth. Medical students working in the First Ward transport these infectious materials from the autopsy room to the First Maternity Ward. The midwives of the Second Ward do not access the autopsy room and therefore are not vehicles of the infection.

To test this hypothesis, Semmelweis required medical students to wash their hands thoroughly with a solution of lime chloride before visiting women in labor.[1] Following this, he was able to find that:

> E6: After the students were forced to wash their hands with a solution of lime chloride, the mortality rate in the FMW dropped to converge with that of the SMW.

E6 confirmed H7 because it provided new evidence that only H7 could explain, and which instead remained unexplained if the other hypotheses were assumed. Furthermore, in addition to E6, H7 is compatible with the remaining available evidence (i.e. with E1-E5), while none of the other hypotheses proposed is. Semmelweis was the first to understand the dynamics through which puerperal fever arises and was a great precursor to the discovery of the role that microorganisms play in infections. His investigation was the first to emphasize these dynamics, and to suggest disinfection as a preventive sanitary means. Ironically, Semmelweis' research was discredited by the medicine of the time and he was fired from the Vienna hospital, despite the positive results achieved, for having given orders without having the authority. His view was confirmed only twenty years later with Pasteur's demonstration of bacterial contamination.

---

1. If it seems absurd to you that those who worked in hospitals did not wash their hands thoroughly, consider that at the time there was still no idea of the need for disinfection in certain environments.

8.3 Abductive procedures

The procedure followed by Semmelweis involved continuously skipping between the evidence and the hypotheses. The evidence led to the formulation of hypotheses that could explain it. This procedure, which goes from evidence to hypotheses, is called, as we said, *abduction*. On the other hand, to test a hypothesis and to compare it with others, the reverse procedure is carried out, that is, from the hypotheses to their consequences. In particular, the fact that a hypothesis has certain consequences allows us to make predictions based on that hypothesis. Such predictions can be disproved or confirmed by the facts. For example, Semmelweis's hypothesis predicted that the mortality rates in the two wards would become the same if the students washed their hands. The various hypotheses and their consequences must therefore be compared with evidence and facts to establish how good a hypothesis is. The case of Semmelweis, as presented, is rather simple because the hypothesis formulated by the Hungarian doctor was evidently better than all the others, as it explained all the evidence and correctly predicted new data. But things are usually not that simple.[2] In many cases, some hypotheses explain some evidence better and other hypotheses other evidence. The hypotheses must therefore be compared with each other and weighed by establishing which one is best on the basis of a plurality of criteria. This procedure is *fallible* because new evidence can always change things and a hypothesis that seems the best may not turn out to be so. It is therefore a form of reasoning that is carried out starting from the evidence which is *contextually available*. This is a crux of inference to the best explanation: we are never guaranteed that it will be conclusive. New evidence could be introduced later, and drastically challenge what had hitherto proven to be the best explanation. Let us now discuss some aspects of abduction that are important for what we say in this chapter. After that, we will move on to discussing the process of *inference to the best explanation*.

8.3.1 Abduction

An inference is abductive if (and only if) it is not deductively valid and its conclusions provide an *explanation* of the evidence that figures in the premises. One of our requirements is that the conclusion of such an argument be a *plausible* explanation. Typically, we use abductive reasoning when we face phenomena or events for which we have no known explanation.

---

2. Semmelweis's case itself has been simplified here and things were actually more complex, so much so that it took twenty years for his hypothesis to be accepted by the scientific community of the time. In particular, Semmelweis believed that the origin of infected materials was exclusively cadaverous, while in fact this is not the case.

Example (Lost Smartphone). While you are leaving a university classroom with a colleague, he realizes that he does not have his smartphone with him. You saw him with his smartphone inside the classroom during the lesson just concluded. Then you tell him: "You left your smartphone in the classroom".

This is an extremely simple example of abductive reasoning. Let us briefly reconstruct it. You used the evidence expressed by "My colleague lost his smartphone", and "My colleague had his smartphone in the classroom we just left" as the basis for a reasoning process aimed at finding out why your colleague no longer had it, and that evidence led you to conclude the hypothesis expressed by "My colleague left his smartphone in the classroom we just left". In addition to explaining why he could not find it on his way out, the hypothesis *agrees* with the evidence available – and in particular, with the fact that he had his smartphone while he was in the classroom.

Abductive reasoning operates "from evidence to hypothesis". It starts from some data (evidence) using them as *premises*, and proceeds to propose an *explanatory hypothesis* as its own conclusion. Abductive reasoning aims to conclude an explanatory hypothesis that is *plausible* given the evidence we have on the phenomenon we wish to explain.

If we look at things from a *deductive* point of view, abductive reasoning is incorrect, because it commits the fallacy of affirming the consequent – see Chapter 6 on this. Take the smartphone example. It corresponds to this reasoning pattern:

(E1) My colleague cannot find his smartphone.
(E2) If my colleague has left his smartphone in the classroom, then he cannot find it.
___
(H1) My colleague left his smartphone in the classroom.

Note that *we cannot validly deduce the last sentence from the first two*. To realize this, consider that there are explanatory hypotheses which are compatible with the first two sentences but different from (H1). For example, knowing that my colleague is very distracted (E3) and has a bag full of things (E4), I could conjecture that he put his smartphone in his bag without even noticing, but now, even by rummaging there, he still cannot find it (H2). It is easy to see that (E3) and (E4) are compatible with (E1) and (E2), but if added to them lead us to conclude (H2) rather than (H1).

How then should we deal with *abductive inference*? Simply, as a form of reasoning that allows us to formulate a hypothesis in response to a set of starting evidentiary items. If we try to use it as a tool for inferring conclusions in a deductively valid or conclusive way, then we will make the logical mistake of treating abductive reasoning as if it were deductive.

Note that abductive reasoning is, like induction, *ampliative*, *fallible*, and *non-monotonic*.³ Let us consider it using the smartphone example. As regards the character of being *ampliative*, the information provided by (H1) is not implicit or already contained in that provided by (E1) and (E2) taken together, and in the same way, the information provided by (H2) is not contained in that provided by (E1)-(E2) or in that provided by (E1)-(E4).

As to *fallibility*, (H1) may be the most plausible hypothesis given (E1) and (E2), but however good the abductive reasoning in question is, we may find that (E1)-(E2) are true and (H1) is false – for example, (H2) could be true instead of (H1).

Finally, abductive reasoning is *non-monotonic*: (E1) and (E2) lead us to conclude (H1), but if we add (E3) and (E4), we are no longer in a position to conclude (H1) as we did earlier.

The smartphone example shows that the kind of abductive reasoning presented above is useful for the *formulation* of a hypothesis, a key step in any attempt to explain a phenomenon. However, abduction is not useful as a way of *testing* the available hypothesis. In fact, facing a phenomenon to explain, there is usually not just one possible explanatory hypothesis, but two or more *rival* hypotheses to choose from; in our example, they are (H1), namely the hypothesis that the smartphone was left in the classroom, and (H2), the hypothesis that the smartphone is in the bag. Abductive reasonings such as those we have exemplified do not help us to say which hypothesis is more solid. To do this, we must add further steps to the reasoning that leads us to formulate a hypothesis. It is the process of *inference to the best explanation* (IBE).

8.3.2 Inference to the best explanation

The smartphone example exemplifies the stage of *hypothesis formulation*, the one where reasoning goes "from evidence to hypothesis". But when hypotheses have to be evaluated and if possible tested, which includes going "from hypothesis to evidence", we need other procedures.

The *testing of hypotheses* is actually the central aspect in the non-deductive reasoning procedures via which we come to select a hypothesis that explains a given phenomenon. Note that, in the hypothesis testing phase, we also resort to deductive reasoning. Take for example the case in which we add new evidence to the initial one, precisely to test a given hypothesis, and this new evidence contradicts the hypothesis. This is a case of *falsification*. Ideally, what happens in this case is that the hypothesis implies something that is negated by the new evidence.⁴ From this,

---

3. See Chapter 2 on these three concepts, § 2.3.3 in particular.
4. In fact, this process can also involve the initial evidence. Be that as it may, what is implied by a hypothesis must be different from the evidence that suggested the hypothesis, in order to avoid the process being circular.

we conclude the negation of the hypothesis. This type of inference is what is called *modus tollens* (again, see Chapter 6), and it is precisely the kind of deductive reasoning we follow when we falsify a hypothesis. When, on the other hand, the test does not lead to a falsification or to a conclusive verification of it, the hypothesis is *confirmed* or *disconfirmed* – that is, it proves to have more or less solidity, plausibility, probability, etc., without the observers being in a position to conclude *beyond all possibility of revision* that it is true or that it is false. In this case, the inferences involved in moving "from hypothesis to evidence" take the form of complex evaluations that seek to determine the "best explanation" of the available evidence, without this being able to impose itself with absolute certainty. IBE is the type of procedure consisting in this.

When can we qualify a given explanation as *best* (assuming there is a best one)? The criteria that guide a comparative evaluation between hypotheses are, in part, general (epistemic criteria such as plausibility, coherence, completeness, simplicity) and, in part, linked to the context of the explanation (if there are particular requirements or standards in the relevant context).

In the following we mention some general criteria for judging a hypothesis as more or less good and some criteria for comparing hypotheses with one another. As for the general criteria, these are some of the most widely employed:

1. *Internal consistency.* This criterion concerns the ways in which the various parts of the hypothesis are connected to one another. Clearly if a hypothesis contains contradictions, it will not be a good one. But also when a hypothesis provides different explanations for phenomena that are clearly interconnected, we cannot judge that hypothesis as good. For example, in the nineteenth century very different explanations were circulating for magnetism, on the one hand, and electricity, on the other, but when, towards the end of the century, physicist James Clerk Maxwell discovered that the two phenomena are interrelated, these explanations lost their credibility.
2. *External consistency.* An explanatory hypothesis is all the better the more consistent it is with other confirmed facts and hypotheses, as well as with our background knowledge. Suppose, for example, that last night I was woken up by the strong wind and that this morning I found some overturned plants on the terrace. Suppose we formulate three hypotheses: a) aliens landed on my terrace and knocked my plants over just for fun; b) thieves climbed onto my balcony with the intention of breaking in but then did not do so, and when they climbed onto the terrace they bumped into the plants; c) the wind knocked the plants down. Why do we tend to consider c) the best explanation of the three? One of the main reasons for rejecting a) is that it is inconsistent with everything we know and believe is established. As far as we know, aliens do not land on people's balconies and in all likelihood have never landed on Earth (at least for now). Hence a) is not consistent with all the rest of what we know. Even if to

a lesser extent, consistency with all the rest of our knowledge constitutes at the same time one of the reasons for rejecting b) and considering c) the best explanation. Indeed, there are no signs of break-ins in my house, no one has reported thieves, no one has heard any suspicious noise; c) is therefore the hypothesis that best accords with all the rest of our knowledge. This criterion obviously does not imply that it is not possible to put forward new hypotheses that are not consistent with what we already believe. If that were the case, knowledge would never advance. External consistency is only *one* of the criteria available to us. However, a new hypothesis that is not consistent with what we believe must have many merits and must be well confirmed in order to be accepted.

3. *Explanatory power*. A hypothesis is better the greater the number of facts it explains. Those that explain a few facts are not good hypotheses. Those that explain a single fact, called *ad hoc* hypotheses, are generally bad ones. Let us go back to the previous example and suppose that not only do I see the plants on my terrace knocked down but that I also see that some of the plants on the balconies of the building opposite to mine have fallen too, that the trees on the street have lost many leaves and branches, that one of these was rooted out. The strong wind hypothesis not only explains the state of the plants on my terrace, but also the many other facts I observe. Conversely, the thieves hypothesis only explains the state of the plants on my terrace, but not everything else. This is why the wind hypothesis is better. In fact, the comparison between the explanatory power of two theories is often not a simple matter because it is not merely a quantitative question. In fact, not all the facts that a hypothesis explains have the same importance. It is also necessary to evaluate how the hypothesis behaves towards new evidence that could emerge. We could then divide this test into three different sub-tests:

(A) *Quantity test*: if H_1 explains n items of evidence and H_2 explains m items where m is greater than n, then $H_2 > H_1$ (where ">" expresses a preference relation).
(B) *Quality test*: if H_1 explains less important items and H_2 explains more important ones, then $H_2 > H_1$.
(C) *Prediction test*: if H_1 does not successfully predict any new evidentiary item while H_2 does, then $H_2 > H_1$.

More analytically, the Prediction test can be specified as follows:

(C.1) No prediction < confirmed prediction.
(C.2) Falsified prediction < confirmed prediction.
(C.3) No prediction > falsified prediction?

Point (C.3) is uncertain because it is not clear, out of context, whether the condition "no prediction" should be preferred over the condition "falsified prediction". On the one hand, the "falsified prediction" condition usually means that the hypothesis has been falsified. In this sense, "no prediction" is a better condition for the hypothesis. On the other hand, the fact that a prediction has been made and falsified shows that the investigation has advanced. In this sense, the "falsified prediction" condition is better.

The medical investigation at § 8.2 offers us a quite easy case, since H7 beats the other six hypotheses in *all* tests. This fortunate situation does not necessarily occur in every comparison of hypotheses, however. In this regard, we must be aware that it is difficult to establish "second order tests" – that is, tests that tell us which of the different criteria is more important. For example, if a hypothesis is not consistent with our previous assumptions, but is internally consistent and has great explanatory power, how should it be treated? What if a hypothesis does better in quantitative test (A) but a rival scores better in quality test (B)? It is difficult to say in general and out of context which test carries more weight in the comparative assessment of hypotheses.

8.4 An example of reasoning with explanatory hypotheses: reasoning from effects to causes

In this section we present a specific type of IBE, which is involved in causal reasoning, and more precisely in the reasoning that goes "from effects to causes". In many cases, when the explanation for an event, fact or phenomenon is sought, the causes of that fact are sought. Hence, causal explanations are an important type of explanation. But not all explanations are causal. For example, modern physics has explained the electric current by interpreting it as a flow of electrons and yet it does not seem correct to say that a flow of electrons is the cause of the electric current. Rather, from a certain point of view, the electric current and the flow of electrons are the same thing. The contemporary physical theory of electron flow thus provides a non-causal explanation of electric current. While not all explanations are causal, there is no doubt that causal explanations are an important type of explanation, perhaps the most important. A further reason to focus on it is that reasoning from effects to causes is an important daily activity, for example in specific contexts like legal cases. For instance, in a trial a judge may need to establish what caused the victim's death, or determined damage to property. In matters of public interest, we may need to determine what caused a crisis in the subprime investment market, and so on.

Let us start with an example:

Example of toxoplasmosis. In mid-March 1968, five Cornell University medical students fell ill and visited the University's Student Health Service. They had flu-like symptoms, such as headache, fever, and body aches. After a blood test, the doctors found that the students had *toxoplasmosis*. In their daily routine, all five students *attended different class-*

es, lived in different places, barely knew each other, and ate in different places. However, it turned out that all five students had eaten rare burgers at the dormitory snack bar on the night of March 5. In addition, they had only consumed sandwiches, soft drinks and coffee that night. Apparently, the only significant fact was that everyone ate rare burgers in the same place and on the same night. The doctors dealing with the case concluded that the five patients had contracted toxoplasmosis *that* night, and *in the same way*: by eating hamburgers that had not been well cooked. In other words, the ingestion of poorly cooked meat had *caused* the transmission of toxoplasmosis in the five cases considered.

The doctors in the example are engaged in *causal reasoning*. More precisely, they are reasoning *from the effects* (transmission of toxoplasmosis) *to the causes* (ingestion of badly cooked burgers). Moreover, the doctors' conclusion is a *causal statement* (ingestion of undercooked meat caused the transmission), which is a statement of the form "X causes Y". Further, all their premises are factual statements, that is, statements that record a series of relevant facts.

This is typical of our reasoning from effects to causes: we start with the data on facts, and draw a conclusion that connects two or more of these facts in a causal way. In the example described above, the doctors start from data that include "Students ingested undercooked meat" and "Students have toxoplasmosis" and conclude that the former caused the latter. In doing so, they implicitly followed a general method of finding causes known as the "method of agreement" which we will return to shortly.

It is clear that the *example of toxoplasmosis* offers us a case of IBE. Doctors had a number of hypotheses at their disposal, at least ideally. For instance, that the five students had passed toxoplasmosis on to each other (H1); or that they had contracted it in common environments frequented in their daily routine (H2); or that they had contracted it from eating undercooked meat from the same batch (H3). H1 and H2 do not agree with the fact that the five students did not hang out with each other (E1) and that they did not share any of the spaces of their routine (E2), including the places where they usually ate. E2 also seems to disprove H3, but note that the five students ate in the same place on the night of March 5, 1968 (E3) and all of them ate meat (E4), and nothing else that can carry toxoplasmosis. E3 and E4 agree with H3, and therefore suggest that this is the best of the possible hypotheses. In other words, H3 explains the toxoplasmosis transmission and is the only one that agrees with the evidence. As a result, it is the best explanation of the case.

Logical research into reasoning from effects to causes has systematized five rules that are, under certain circumstances, solid and reliable. These rules are known as the "method of agreement", the "method of difference", the "joint method of agreement and difference", the "method of residue", and the "method of concomitant variations". We owe these to the British philosopher John Stuart Mill (1806-1873), who dealt with topics ranging from logic and philosophy of language to economics and political philosophy. These five methods are conclusive in eliminating hypotheses, and can be illuminating in offering positive, albeit tentative, conclusions regarding what causes what.

Mill's methods start from (i) an effect Y whose cause must be identified, and (ii) a number of relevant factors A, B, C, D, ... which could be the cause of Y. Thus, each method proposes a specific test to select one (or at least some) of A, B, C, D, ... as the cause of Y. In what follows, we will introduce the first three methods of Mill, in which the connection with the more general procedure of IBE is clearer. In doing so, we will show precisely how methods are forms of IBE.

Before moving on to Mill's methods, let us briefly discuss three characteristics that are crucial to understanding what these methods allow or do not allow us to do:

1. Mill's methods do not help us discover *new facts*.
2. Mill's methods offer *no conclusive evidence* on causal connections.
3. Mill's methods test our assumptions on *what caused what*.

No new facts discovered. Mill's methods start with known facts (the relevant factors) and link them to the effect to be explained by providing some form of control of that link. No new facts are discovered through the five methods (think of the example of toxoplasmosis). The methods may lead us to look for new facts so that we get more relevant information about what causes what, but they do not allow us to *discover* these new facts.

No conclusive evidence. Mill's methods – like any form of non-deductive reasoning – cannot provide conclusive proof that a certain fact X caused a certain fact Y, but only a tentative explanation, which can be abandoned if new evidence is added. Take the example of toxoplasmosis. If doctors were later informed that all students had been drinking unpasteurized milk, doctors would be prompted to retract the conclusion that meat was the cause of transmission. This is enough to see that reasoning from effects to causes is *non-monotonic*, like all evidence-based reasoning. Likewise, it helps us understand that it is *fallible*. Let us imagine that the transmission was caused by ingestion of milk, but that the students did not remember drinking it and therefore did not tell the doctors. As a result, we would find ourselves in a situation in which a decisive piece of evidence would be unknown. Under these conditions, the causal hypothesis of undercooked meat would be false, but nevertheless the reasoning that leads to it would still be good, as the evidence available to doctors, which does not include the ingestion of milk, provides reasons for it.

A way to check the hypotheses. Ideally, with Mill's methods we select some previous factors because we assume that they are candidates for the causal role with respect to a certain effect Y. For example, in the example of toxoplasmosis, the factors concerning the environments attended, the contacts between the students and food in-

8 Reasoning with explanatory hypotheses

gestion are the previous factors. The tests defined by each of Mill's methods select some of these factors. Hypotheses that resist the check made by the relevant test are considered the cause (or causes) of Y.

8.4.1 The method of agreement

The method of agreement is applied to the toxoplasmosis example in § 8.4, and functions as a test for the necessary conditions of effect Y. X is a *necessary condition* of Y if Y does not occur in the absence of X. The method of agreement can be formulated as follows:

> If in all cases $C_1, ..., C_n$ where an effect Y occurs, there is a single preceding factor X that is shared by all those cases, then X is the cause of Y.

A "preceding factor" is something that happened before the effect occurred in cases $C_1, ..., C_n$. Of course, we want to choose preceding factors that are relevant, so as to avoid introducing misleading elements in our reasoning. The following example illustrates how the method of agreement works:

> Four friends go out for dinner together, but when they get home they all start to feel sick and suffer from stomach pain.

The relevant factors in this case are what the four friends ate or drank for dinner. On the contrary, the clothes they wore are not a relevant factor. Table 8.1 summarizes what each of them ate and notes who got sick:

Table 8.1 The method of agreement

	A (Chicken)	B (Pasta)	C (Red wine)	D (Beer)	Y (Sickness)
Case 1	YES	YES	YES	YES	YES
Case 2	YES	NO	NO	YES	YES
Case 3	YES	YES	NO	NO	YES
Case 4	YES	NO	YES	NO	YES

Chicken is the only thing all four friends ate or drank that night. Applying the method of agreement, we conclude that chicken is the cause of the stomach pain.

Using the method of agreement, all steps of the IBE are at work. In particular, the information on the dishes that have been eaten allows us to formulate four different causal hypotheses ("It was the chicken that caused the stomach pain", "It was the pasta that caused the stomach pain", etc.). The hypotheses are compared with the evidence ("Everyone was sick", "Everyone ate chicken", "Not everyone ate pasta", and so on). For each hypothesis, we note when the relevant factor correlates with stomach pain, and when it does not. The four hypotheses are then compared with one another regarding their correlation with stomach pain. What relevant factor is present in all cases of stomach pain? Ingestion of chicken. What is required for the causal explanation? That it explains all the evidence. Which hypothesis does that? The one that it was the chicken that caused the stomachache. It is concluded that the best hypothesis is this: "It was the chicken that caused the stomachache".

8.4.2 The method of difference

The method of difference is a complex test that calls into question the notion of *sufficient condition*. X is a sufficient condition of Y if the occurrence of X is sufficient to determine the occurrence of Y. The method can be summarized as follows:

> Given the cases $C_1, ..., C_n$, if an effect Y occurs in all cases except C_i, and all cases share all factors except C_i, then the factor that only C_i does not share is the cause by Y.

The method can be seen as the following list of instructions: (1) consider both the cases in which effect Y occurs and those in which it does not; (2) eliminate all the factors which are present even when Y does not occur and which are therefore not *sufficient conditions* for the occurrence of Y; (3) if there is a factor that we cannot discard because it is not present when Y is not present, that factor constitutes a *sufficient condition* of Y and therefore can be considered its cause. Let us adapt the example above, in order to have a concrete situation where this method can be put to work.

> Four friends go out for dinner together, but when they get home, they all begin to feel sick and suffer from stomach pain, except one of them.

The cases and preceding factors of the new example are these (Table 8.2):

Table 8.2 The method of difference

	A (Chicken)	B (Pasta)	C (Red wine)	D (Beer)	Y (Sickness)
Case 1	YES	YES	YES	YES	YES
Case 2	YES	YES	YES	YES	YES
Case 3	YES	YES	YES	YES	YES
Case 4	YES	YES	NO	YES	NO

In this case, the fourth friend is the only one who does not get sick and the only difference between him and the others is that he did not have red wine (C). Chicken, pasta and beer are present even when the disease is not present, and therefore do not constitute a sufficient condition for the sickness effect. This leaves red wine

8 Reasoning with explanatory hypotheses

alone as a candidate to be the sufficient condition of the effect as, *when it is lacking, the effect is also lacking*. Hence, the method of difference makes us conclude that red wine is the cause of the sickness.

The difference method is also an example of IBE. Once again, the information on the food and beverages that have been ingested allows us to formulate four different causal hypotheses. The hypotheses are compared with the evidence ("All those who got sick ate chicken and pasta, and drank red wine and beer", "All those who didn't get sick did not drink red wine"). This leads us to select the hypothesis that the best candidate for the cause of the sickness is red wine.

8.4.3 The joint method of agreement and difference

The joint method (of agreement and difference) does nothing but apply agreement and difference at the same time, thus guaranteeing a more rigorous screening and identifying the necessary and sufficient conditions for the occurrence of given effects. The method is as follows:

> Given the cases C_1, ..., C_n, if there is a factor X present in all cases in which Y is present (while other factors are sometimes absent) and X is absent in all cases in which Y is absent (while others factors are present), then X is the cause of Y.

Another variation of the dinner example will help us understand how the joint method works. Consider the following Table 8.3:

Table 8.3 The joint method

	A (Chicken)	B (Pasta)	C (Red wine)	D (Beer)	Y (Sickness)
Case 1	YES	YES	YES	YES	YES
Case 2	YES	YES	NO	YES	YES
Case 3	YES	YES	YES	NO	YES
Case 4	YES	NO	NO	YES	NO

The joint method suggests that pasta (B) is the cause of stomach pain. In fact, all the friends who got sick ate pasta, and otherwise, none of them ate or drank the same thing. At the same time, the only friend who did not get sick did not eat pasta. In short: pasta alone is not excluded either from the agreement method, if we limit it to the cases in which the effect has occurred, or from the difference method, considering all the available cases.

8.4.4 A more complex case

The three previous examples are very simple, since in each of them only one of the hypotheses passes the test. But what if *more than one* hypothesis passes the test? Let us consider this modified version of the dinner example (Table 8.4):

Table 8.4 Complex case

	A (Chicken)	B (Pasta)	C (Red wine)	D (Beer)	E (Oysters)	Y (Sickness)
Case 1	YES	YES	YES	YES	YES	YES
Case 2	YES	NO	NO	YES	YES	YES
Case 3	YES	YES	NO	NO	YES	YES
Case 4	YES	NO	YES	NO	YES	YES

With the information now available, it would be wrong to conclude that chicken caused the disease: all sick friends had also taken oysters, and we are not able to choose between chicken and oysters. In this case, all we can conclude is that the combination of chicken and oysters caused the disease, or that either chicken or oysters caused it, but we do not have enough data to say which of the two hypotheses is correct. This example can also help us understand some other features of Mill's methods.

First, suppose we were initially only informed of the first four factors and that it was *only later* that we learned that our friends had eaten oysters. We would first conclude that chicken was the cause of the disease, and we would later drop this conclusion. This shows that conclusions about causal connections can be dropped and replaced if new information is available (*non-monotonicity* and *fallibility*).

Second, the example helps to understand that, in the absence of complete information (which is the usual condition), our causal conclusions are only *provisional*.

Finally, all we can conclude from the information available in the new example is that oysters, or chicken, or their combination, caused the disease. We may want to know which of the three options is the correct one. To do this, we need to get more information. In this case, the natural thing to do is to get information about other clients who suffered from stomach pain that night. If at least one of them ate only chicken (or only oysters), we would be justified in concluding that the chicken (or oysters) caused the disease.[5] If all of them ate both chicken and oysters, we will have to settle for the weaker conclusion that the combination of the two caused the stomachache (the additional information we find may not be decisive, sometimes). Of course, the increase in information can also lead us to eliminate some of the candidates we had taken for granted. For example, if any customer has suffered

---

5. Actually, to conclude this way, we also need to know that none of the other clients ate only oysters (or only chicken).

from stomach pain without having eaten chicken or oysters, we should conclude that neither chicken nor oysters caused the sickness.

8.4.5 An assessment of Mill's methods

We conclude by pointing out some limitations of the reasoning procedures from effects to causes systematized by Mill.

Selection of relevant preceding factors. The first step in applying Mill's methods is the selection of the relevant preceding factors, and this is by no means an easy matter, unless the case in question is very simple. Take this example:

> *Drunk alien.* An alien lands on Earth. He does not know what alcohol is, but he observes that humans are intoxicated by many different drinks. He wants to understand what causes the intoxication, and he drinks for five evenings. He notes carefully what he drinks in the five evenings: whisky and soda, brandy and soda, rum and soda, gin and soda, bourbon and soda. He concludes that it is the *soda* that causes his intoxication.

The alien is following the agreement method and does it correctly, but comes to a wrong conclusion: soda does not cause drunkenness, of course. This happens because the alien does not know that there is a relevant factor (alcohol) that is shared by all the drinks he has drunk, just like soda. Had he known, he would have come to a different conclusion. More generally, the alien does not know what kinds of things may be relevant to his inquiry.

In general, when we are well supported by a good background knowledge of relevant factors (as in the example of toxoplasmosis), Mill's methods are quite robust. Poor background knowledge, on the other hand, makes Mill's methods less effective. This somewhat limits their application to a precise set of cases: those in which we have a clear idea of the possible causes and just want to check which of them was actually at work.

Indeterministic causality. Mill's methods work quite effectively for those cases where causality is *deterministic*. These are cases where the effect occurs with certainty if the cause occurs. The dinner examples presented above fall into this category. However, many cases of causation are *indeterministic* and *probabilistic*. Throwing a stone at a window can cause it to break, but this does not always happen. A heart attack can result in death, but again this does not happen all the times. Driving recklessly increases the likelihood of causing accidents but there is no certainty that this will occur. In all these cases, the occurrence of the cause increases the likelihood of the effect, but does not determine the occurrence of the effect.

Mill's methods can give wrong results when dealing with probabilistic causes. Take the joint method, for example, where the cause is supposed to be a condition that occurs if (and only if) the effect is present. This excludes causes which do not determine their effects but which make them "only" highly probable. For example, suppose that in the example of the dinner in § 8.4.3 the fourth friend also ate pasta, but did not feel sick. The joint method would rule out pasta as a cause of sickness, but perhaps pasta considerably increased the probability of being sick, and the fourth friend escaped stomach pain just because he may have had great resistance to the element that caused the intoxication in the other friends.

This is a considerable limitation, as a good number of interesting causal connections are probabilistic. A paradigmatic example: Researchers argue that smoking is one of the possible causes of cancer, as there are a number of tested connections between smokers and cancer, and these connections include the fact that smoking dramatically increases the chances of contracting cancer. However, not all people who smoke (even heavily) get cancer. Smoking is therefore only a probabilistic and nondeterministic cause of cancer.

8.5 Conclusion

Let us briefly recap the contents of this chapter. We introduced the notion of *reasoning with explanatory hypotheses*, and hinted at the presence of many different models for the notion of explanation. In particular we have distinguished between *deductive* models of explanation and *non-deductive* models, and we have concentrated on the latter. We discussed an actual case of reasoning with explanatory hypotheses, analyzing the explanation of the causes of puerperal fever given by the Hungarian doctor Semmelweis. The example allowed us to see what the functions of evidence are in suggesting and testing hypotheses in practical terms, and how a given evidentiary item can falsify or confirm a hypothesis. We then moved on to articulate a framework of notions and tools that would help us understand the general aspects and dynamics of reasoning with explanatory hypotheses. In particular, we discussed abductive reasoning and inference to the best explanation, defining their various steps and then introducing the general criteria for evaluating the hypotheses not only in relation to the evidence but also with regard to their rival fellows. We discussed the criteria of internal consistency, external consistency and explanatory power. The latter has made evident the need for *tests* against which to compare rival hypotheses. We have seen that there is a plurality of such tests, and we have mentioned the crucial fact that, when faced with this plurality, it is not easy to define second-order tests, i.e. tests that tell us which of the various tests should be considered more important than the others. Finally, we discussed three of Mill's five methods as specific cases of inference to the best explanation. More precisely, we have seen how such methods help us compare *rival causal hypotheses* – where these are hypotheses about what causes a given effect. We closed the chapter by discussing the limitations of Mill's methods, which emerge in relation to nondeterministic causality and in the presence of a causal investigation that does not have an adequate background knowledge supporting it.

9 | Statistical reasoning

In this chapter we will deal with statistical generalizations. We will distinguish two types: purely statistical generalizations (§ 9.2) and generalizations concerning the future (§ 9.3). We will also consider an *application* of statistics to particular cases through an inference scheme called "statistical syllogism" (§ 9.4). In addressing these issues, we will introduce some basic notions that are relevant in statistical reasoning, for example, *sample, target property, measured property, accuracy* and *margin of error*, and we will see how these concepts play an important role in assessing the strength of a statistical generalization or a statistical syllogism. We will then discuss properties such as the *significance* and *accuracy* of the measured property against the target property, the *size suitability* and *representativeness* of the sample, the possibility that a question is biased, and that *hidden variables* mislead us in our considerations. We will also briefly discuss the significance of the margin of error in our statistical generalizations. We will relate generalizations about the future to scientific laws and see why these generalizations are weaker than those called "purely statistical". In connection with statistical syllogisms, we will finally see the importance of choosing the right *reference classes*.

9.1 Introduction

A *generalization* is an argument that (i) starts out from the fact that all or a percentage *n* of individuals in a *sample* have a property *P,* and (ii) concludes that *all* or *n%* of the individuals in the *population* from which the sample is taken have property *P*. We use generalizations very often in everyday life and they are fundamental in order to interact advantageously with our environment. When we were very young children, we observed to our own expense that we hurt ourselves when we fell on hard ground. We concluded that *every time* (or at least the vast majority of times) we fall on hard ground, we get hurt. We have therefore generalized from the observation of some cases – those in which we injured ourselves by falling – to all cases in which we fall on hard ground. Consequently, in principle, we try to avoid falling when we are on hard ground. Generalizations are therefore fundamental as they help us survive in an environment that can cause us harm and negative effects. Statistics is a science that helps us make our generalizations more rigorous and precise.

A well-known aphorism by Mark Twain states: "All generalizations are false". This statement cannot be true, of course: it is itself a generalization, and therefore if it were true, it would also be false. Twain's aphorism, however, highlights the fact that statistics are a type of *inductive reasoning*, and as we have already said in Chapter 2, *in inductive reasoning the conclusion does not logically follow from the premises*. Even if the premises are true, the conclusion may turn out to be false.

We can consider two types of statistical generalizations:

1. *Purely statistical generalizations.* These are generalizations that depart from a sample and from the observation that a certain percentage of the sample has property *P*. They conclude that *the same percentage* of the entire population of which that sample is a subset has *that property.*
2. *Generalizations concerning the future.* These are generalizations in which the *sample* includes only cases that are already given (what we will call "past cases"). However, the *population* of which the sample is a subset also includes cases that have not yet been given (what we will call "future cases"). In this case, the reasoning is not purely statistical but is based on the additional premise that the population and its distribution of properties (including the property *P* we are interested in) are uniform over time. As we will see, this premise is often not trivial.

Let us now consider these two types of generalizations in detail.

9.2 Reasoning with purely statistical generalizations

Let us start with an example. Imagine that an opinion poll is performed to determine which candidate for a certain political office is the most popular with public opinion. Researchers interview 1,000 people and ask them whether they like candidate *x*, candidate *y*, candidate *z*, and so on for all candidates. Suppose 371 subjects report that they like candidate *x* and that this number is greater than that given for all other candidates. The researchers conclude that candidate *x* is *the most popular* with the public.

The basis of inductive reasoning – in our example, the 1,000 respondents – is called the *sample*, and it is a subset of the *population* which we want to know about. In this example, the population is the totality of those who have the right to vote in the general elections. Since the sample represents only a subset of the population, the conclusion that candidate *x* is the most popular does not logically follow from the premises. It is a *projection* from the sample result. It is clear why this term is used: the conclusion *projects* the distribution of properties *from* the sample *over* the entire population. The property under investigation – in the example, being the most liked candidate – is called the *target property*. Note that we often do not directly investigate the target property. In our example we do not directly investigate the property of "being the most popular candidate"; instead we investigate the property of "being the candidate who is said to be most popular". In other words, we are not really looking into *what preferences interviewees have*, but *what preferences interviewees say they have*. The reason is clear: in this case we do not have access to the target property (the preferences of the respondents), but we do have access to a property that is somehow related (what the respondents report about their preferences). The property that we check to get an idea of the distribution of the target property is called the *measured property*. Obviously, the measured property must have a *meaningful* connection with the target property. This is not a trivial point, even in cases where we take such a connection for granted. We will return to this point shortly. We say that the measured property is more or less *accurate* to indicate how good an indicator of the target property the measured property is. Since generalizations are inductive reasoning, they always have a *margin of error*, even if it is often not reported. The margin of error tells us what "divergence from statistical generalization" we can expect in a given case. For example, suppose the margin of error of the statistical generalization in our example is 2%. The generalization concludes that 37% of voters prefer candidate x. Let us now assume, for simplicity, that *all* eligible voters vote. If the margin of error is 2%, we consider the statistical generalization still "satisfactory" if the candidate receives 35% or 39% of the preferences, but not if she receives 34% or 40% of the preferences.

The scheme of inference of *purely statistical* generalizations is the following:

n% of the sample has the measured property Q.
---
n% of the entire population has the target property P.

Statistical generalizations are not procedures of deductive reasoning, but this does not mean that they are arbitrary. They are justified under a mathematical theorem according to which, given a set S and assuming that a percentage x of the members of S have property P, then *most* subsets of S with sufficiently large cardinality have a similar percentage of members who possess property P. This theorem underlies the reliability of statistical generalizations. However, such generalizations, like all inductive reasonings, can be strong or weak. Below we will look at some ways in which arguments containing statistical generalizations can turn out to be bad arguments.

9.2.1 Lack of relationship between the target and measured property

A statistical generalization will be bad, of course, if the proper relationships between the *measured* and the *target* property are missing. Taking an extreme example: if I concluded that 37% of voters prefer candidate x *on the basis* that 37% of the sample prefer one car model, my statistical generalization would have no value. What people say they prefer about cars is *not* an indicator of what they prefer about political candidates. The two properties lack *relevant* relationships.

There are less extreme cases where there is a significant relationship between the two properties, but nevertheless the measured property is not accurate – or at least, it is *less accurate* than we would expect. For example, the target property of the previous case is the approval rating of candidates, while the measured property is the answers given by interviewees. There is obviously a significant connection between the two, but it has been demonstrated that, often, interviewees do not give honest answers to interviewers: for example, they say they do not like an embarrassing candidate but then end up voting for that very candidate in the privacy of the voting booth. Or they try to present themselves in a better light, even though they know the questionnaire is anonymous. The relationship between target property and measured property must therefore be carefully evaluated so that the generalization is not undermined. Since accuracy is a property that varies on a continuous scale, the same also happens with statistical generalizations: the more accurate the measured property is with respect to the target property, the stronger the generalization will be, excluding other problems that we will look at shortly. Conversely, the less accurate the measured property is, the weaker the generalization will be.

9.2.2 Cardinality of the sample

The XKCD cartoon (Figure 9.1) shows that from an obviously inadequate sample (today, her wedding day), an obviously inadequate conclusion can be drawn (by the end of the month she will be married a lot of times). In general, when we are evaluating an argument containing statistics, the size of the sample matters. In particular, it is mandatory to evaluate the relationship between the cardinality of the sample and that of the population under examination. As we have seen, the mathematical theorem on which statistical generalizations are grounded states that, to be representative of S, a subset of S' must have a sufficiently large cardinality. Too small a sample is likely not to be significant because it may have a distribution of properties very different from that of the entire S.

Consequently, a statistical generalization based on a sample of *inadequate size* will be bad. Therefore, the following reasoning

> 100% of a sample consisting of a single day has the property that a given girl marries on that day.
> _____
> 100% of the days have the property that that girl marries on such days.

is a *bad* statistical generalization. In this case we speak of *unwarranted generalization* (cf. § 4.2.2).

The cardinality of the sample affects the margin of error – which indicates how much the sample result may diverge from the population result. Inferring from a sample of twenty Italians that Italians (do not) know the name of the first President of the Italian Republic is generally less correct than inferring the same con-

Figure 9.1 Cardinality too small

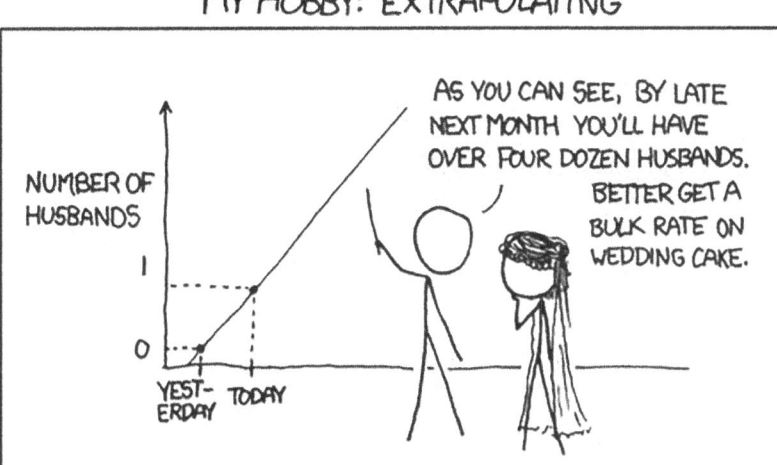

Source: https://xkcd.com/605/.

clusion from a sample of 2,000 people. *Generally, the larger the sample, the smaller the margin of error of our result.* This is how we reason when we choose a restaurant through a social network: the higher the number of reviews, the more we trust their average.

One might think that the larger the population, the larger the sample size should be. However, that is not exactly the case. The relationship between the size of the sample needed to have good quality generalizations and the size of the population is not constant. When the population is very large (more than 100,000 individuals, for example) it is no longer necessary to increase the size of the sample as the population increases. For example, for a population of 10 million individuals it is sufficient to have a sample of about 1,000 individuals to obtain results with a margin of error of 3%. This is roughly the same sample size we need for a population 100 times smaller.

9.2.3 Biased samples

Sample size is not the only property that affects the strength of generalizations. Another crucial property is the *representativeness* of the sample with respect to the entire population. Inferring that all Italian adults know the name of their first President from a sample of 2,500 scholars in contemporary history would not be correct, even though the sample is sufficiently large. This is because we are relying on a subset of the population in which there is an extremely high distribution of the property of "knowing who the first President of the Italian Republic was"– indeed, *one much higher than the average distribution* in all Italian adults. In fact, it is reasonable to expect that the distribution of the property in the sample would be different from the mean distribution in the population. The sample we have chosen, in short, is *not representative* but is biased, and the statistical generalization is therefore bad. More specifically, the following generalization is bad:

100% of the sample of contemporary historians know who the first President of the Italian Republic was.

100% of Italian adults know who the first President of the Italian Republic was.

Choosing a representative, unbiased sample is not easy, and choosing at random does not always work: not even randomizing is easy, although this may sound odd. Let us examine an example from real life. In 1936 the magazine *Literary Digest* sent out 10 million questionnaires asking which candidate the recipient would vote for in the next American presidential election: Franklin Roosevelt or Alf Landon. It received 2.5 million returns and 57.1% of the responses indicated Landon as the favorite candidate. In the light of such a clear majority (far beyond the margin of error) and of the large size of the sample, the magazine confidently predicted that Landon would win by a landslide majority. When the election results came in, Roosevelt had won by an even larger majority in the opposite direction: 62 percent for Roosevelt to a mere 38 percent for Landon. As a result of this misprediction and of its subsequent bad reputation, the *Literary Digest* went bankrupt two years later.

As later studies showed, unexpectedly, the mistake made by the *Literary Digest* was precisely in the representativeness of the sample. The recipients of the questionnaires were chosen *at random* from the telephone directory. The fact that they were chosen at random seemed to ensure that the sample was not biased. At the time, however, there were still few Americans who had telephones, and these were generally owned by city dwellers and by wealthy classes. In this group of people, a large percentage sympathized with Republican ideas. They massively indicated Landon as their preference. That sample was, in short, biased, because the intersection between the set of people who owned a phone and the set of people who voted Republican was large enough to give an above average distribution of preferences for Landon. Those who were poor or who were employed in agricultural jobs outside the cities – people who generally did not have telephones – constituted a larger group who, worn down by the Great Depression, massively voted for Roosevelt.

9.2.4 Question biases

A statistical argument can be biased in other ways. For instance, in opinion polls the way in which questions are asked is very important. The *National Rifle Association* (NRA), to demonstrate that Americans do not wish to limit the possession of weapons, might ask whether the interviewed persons would like *to preserve the constitutional right to keep and bear firearms* or *to leave helpless citizens at the mercy of criminals*. Clearly, even if the sample is not biased, the way in which the question is asked is.

Such a question is biased for at least two reasons. In the first place, it foreshadows the danger of a violation of fundamental social rules, thus triggering the same mechanism of *risk aversion* as that discussed in the Asian disease case in Chapter 1. In these cases, we would tend to choose the option *contrary* to the one that exposes us to risk – namely protecting the constitution regardless of any other consideration. Second, the question does not investigate our opinion on the restriction of the right to sell and buy weapons but asks whether we are willing to expose ourselves to violent crimes. However, the generalization from (say) "90% of respondents do not want to expose themselves and others to violent crime" to "90% of Americans want to keep firearms regulation as it is" is undoubtedly a bad generalization. It is obvious that the inference:

n% of respondents do not want to expose themselves and others to violent crimes.

---

n% of Americans want to keep firearms regulation as it is.

is a bad statistical generalization.

Note that this example also raises an obvious issue of relevance: the survey measures the properties "willingness to defend the constitution" and "willingness to expose oneself and others to violent crimes". It is quite clear that there is no significant connection between these properties and the target *declared* property, namely Americans' opinion regarding the restriction on the sale of firearms.

Again, the questions may be biased in less obvious ways. For example, suppose a door-to-door survey is conducted to find out how many people believe in God and go to church. If the survey were conducted by a priest in religious clothing, we would expect this to affect the result: likely, more people would respond in the affirmative than would do so if the survey were conducted by laymen. Therefore, it is often important that respondents do not know the persons conducting the test and whether they are associated with a group, company, political party, etc.

9.2.5 Hidden variables

Consider this scenario. John's grandmother has been in poor health for a while, before being notified that she has to go to hospital for surgery. John has to choose between Hospital A and Hospital B. He is told that in Hospital A, 900 patients out of 1,000 survived surgery, and in Hospital B 800 out of 1,000 did. John is initially inclined to choose Hospital A. Before deciding, however, John stops and considers one crucial property of his grandmother: she is in *poor health*. He then asks about the health status of patients that had undergone that surgery in each of hospitals A and B. In hospital A, 100 patients undergoing that surgery were in poor health. Of those, 30 survived. In hospital B, 400 patients undergoing that surgery were in poor health. 52% of them survived. Once this is taken into account, Hospital A is no longer the obvious choice. In fact, since John's grandmother is in poor health, Hospital B turns out to be better.

The problem here is not that John's initial generalization was bad, but that John was in danger of misusing it. If we do not identify the truly relevant sample – in our case, the set of patients with serious health conditions who undergo the operation – we risk making bad decisions, since they are based on a generalization different from the one we should be considering. This is not a trivial question because there are cases in which we may not have clear ideas about which sample is really relevant.

This problem is known as *the hidden variable problem*, for obvious reasons. In our example, the *variable* of health conditions *was lurking* behind the initial result. This example shows how statistical results can change if the same data are grouped in different ways. It is sometimes said that statisticians can come up with any conclusion whatsoever when given a bunch of data. This is not literally true, but it reminds us to be on the lookout for lurking variables, when reading arguments containing statistics.

In short, there are several factors we should consider when evaluating a statistical generalization. The stronger the connection between the measured property and the target property a generalization is based on, the larger and less biased the sample used and the better the evaluation of the existence of possible hidden variables, the stronger the argument is. Conversely, each of these factors can significantly weaken our reasoning.

9.2.6 Some considerations on the margin of error

Another element we must consider in evaluating a statistical generalization is the margin of error. A *high* margin of error makes the inference *stronger*, but the conclusion less *informative*. Let us examine this, sticking to the example of the popularity of political candidates. Candidate *x* received 37% of preferences from respondents. We concluded that 37% of the voters prefer *x* to other candidates. We also added that statistics have a margin of error. If the election result diverges from the generalization "within" the margin of error, we are led to accept the generalization anyway. Otherwise, we do not accept it and we even find it misleading. Now suppose that the generalization in question tolerates a margin of error of 10%. It follows that the class of electoral results that make it unacceptable is relatively small. In fact, *all* results for which x is between 27% and 47% (a large slice of possible results) make the generalization satisfactory. The inference is more *robust* as it is undermined by a relatively small number of possible scenarios. However, the conclusion is not very *informative*, precisely because it is compatible with a relatively high number of possible scenarios – in short, it excludes fewer possibilities.

Conversely, a *low* margin of error makes the inference *weaker* but the conclusion *more informative*. Suppose we leave everything as it is in the example just proposed, but that the expected margin of error is 2%. In this case, the class of electoral results that make the generalization unacceptable is very large. In fact, *only* the values for x between 35% and 39% (a small slice of possible results) make the generalization satisfactory. The inference is more *fragile* as it is undermined by a relatively large number of possible scenarios. However, the conclusion is more *informative*, precisely because it is compatible with a relatively small number of possible scenarios – in short, it excludes many possibilities.

9.2.7 Statistical generalizations and daily life

Even if we often do not realize it, in our daily lives we often make statistical generalizations and often make the same mistakes as those considered in the previous sections. For example, we have met two dozen French people in our lifetime and because we found them nice, we conclude that all French people are nice, not realizing that the sample is too small and probably biased too (we have probably only met French of a certain type). From the fact that our friends have a certain opinion on a certain issue, we conclude that Italians, or Europeans, or people of a certain sex, generally have that opinion on that matter, not realizing how small the sample of our friends is and how biased it is. We make the same mistakes in interactions with people. For example, when deciding on a person's character, we often rely on the very few interactions we have had, not realizing that that person's behavior in those interactions constitutes too small a sample for evaluating the character of a person.

9.3 Generalizations concerning the future

As we said at the beginning of this chapter, generalizations about the future start out from a given sample of cases, in which property P has a certain distribution, and draw conclusions with regard to a population that is not yet entirely given.

These kinds of statistical generalizations are *less strong* than purely statistical generalizations. In fact, as said, for a generalization to be strong, the sample must be unbiased. But in the case of generalizations concerning future events or facts, a sample that includes *only* past cases could be biased and therefore it cannot be said *a priori* whether it is representative. To understand this concept, take the example of a football team, Ambrosiana FC, which, in the middle of the championship, had played 19 and won 10 matches. Can we generalize and say that at the end of the championship Ambrosiana FC will have won just over half of the matches played? Such a generalization assumes that, in the second part of the championship, Ambrosiana FC will meet teams at the same level as they did in the first part of the championship; that its players will maintain the same physical condition they have kept until now, etc. In other words, a generalization of this type presupposes a *uniformity between past and future* and very often, as in this example, this is not a trivial presupposition. If there is no uniformity between past and future, our sample, including only past cases, will be biased. And since nothing can guarantee such uniformity, generalizations about the future are systematically weaker than purely statistical generalizations. In any case, generalizations about the future are also "strong" or "weak" to varying degrees, and this depends on several factors, such as the number of future "variables" to be taken into account and our ability to ascertain the occurrence of conditions that determine certain future outcomes. For example, the inference

> Ambrosiana FC has so far won 10 matches out of 19.
> ___
> Ambrosiana FC will win 20 matches out of 38 over the entire championship.

is weak, because the performances of Ambrosiana FC are determined by a number of variables that we may not be able to establish and that, perhaps, cannot be established. Think of the muscular and athletic conditions of the players of the team and of other teams, the occurrence of injuries, whether or not other teams will continue to be involved in other competitions, and so on.

By contrast,

> So far, the Sun has risen every day.
> ___
> The Sun will rise tomorrow.

is a much stronger inference, because the variables that *today* determine the position that the Sun will have *tomorrow* are well known thanks to our astronomical theories. We also have considerable confidence in the fact that the factors determining the rising of the Sun tomorrow remain uniform over time, that is, that the physical laws of the motion of the planets are constant and do not change from day to day.

This example also allows us to see that sometimes some generalizations about the future are indirectly based on other generalizations about the future. For example, thanks to our observations we know that the planets of the solar system move in a certain way, in accordance with the Copernican theory subsequently revised by Kepler and Galileo. On the basis of this, we have concluded that *they will move* in the *same way* in the future, and we are thus able to foresee and explain their movements. However, this in itself does not weaken the "dependent" generalizations, such as that about the rising of the Sun, but shows that, often, there is a chain of generalizations that depend on one another. Therefore, even though moving from one generalization on to another can be deductive, one of the premises of this reasoning is obtained inductively. Even the uniformities we feel most confident about cannot be guaranteed deductively. They are hypotheses which, in principle, could be proved wrong in the future.

Scientific laws are generalizations of this kind. They are grounded on the fact that all the samples of a given population have exhibited a certain behavior until now and generalize on future cases: for example, from the fact that in the past we observed that water freezes at 0° at sea level, we conclude that every sample of water freezes at 0° at sea level. We can call this kind of generalizations *universal*: from the observation that a certain property is possessed by 100% of the samples, we draw the conclusion that that property is possessed by all members of the population. Some scholars have argued that universal generalizations concerning future cases, and scientific laws in particular, are a weak form of inductive reasoning because the sample necessarily has a small cardinality compared to the population and because they presuppose the uniformity of nature, which is a hypothesis that cannot be definitely demonstrated.

Be that as it may, economic laws are also statistical generalizations of this kind: they are based on the observation of economic systems and on generalization from past to future cases. The more uniform the behavior of economic systems and agents is over time, the stronger these generalizations become. For example, we can start out with the fact that in the past we observed that a rise in the interest rate produced a reduction in inflation, but also a decline in investment and we can suppose that in the future a rise in the interest rate will produce similar effects. Many decisions taken by economic agents depend on such generalizations.

A particular example of generalization about the future is the following: we observed that all the samples of a certain population have the property P; so, we expect the *next* member of that population to have property P. This kind of inference is called *simple induction* or *simple predictive inference*. It is also very common in our daily lives: for instance, from the fact that every time I ate food x I had digestive problems, I conclude that if I eat the particular instance of x I have on my plate, I will have digestive problems. From the fact that Ann has always responded kindly to my requests, I conclude that she will respond kindly to my next request.

To appreciate the difference between pure statistical generalizations and generalizations about the future, consider the following example:

> All the ravens we have observed so far are black.
> _____
> All ravens that were born and that will be born were and will be black.

This is a generalization regarding the future, because the population onto which I am projecting the property of "being black" (all past, present, and future ravens) does not yet fully exist – only a subset does: past and present ravens. By contrast, the following is a purely statistical generalization:

> All the ravens we have observed so far are black.
> _____
> All the ravens which have existed in the past and which exist today are black.

Here both the sample (the ravens observed so far) and the population (the ravens that have existed in the past and exist today) are fully given. Our scientific generalizations, on the other hand, aim to be generalizations concerning the future, in the light of their *predictive* functions.

9.4 Statistical syllogism

Statistical generalizations go *from* samples *to* populations. But we also use statistics to conclude that *a particular individual* has a property P – in other words, we *apply* statistics. We call this kind of reasoning "statistical syllogism". Let us look at an example.

Suppose we have made a statistical generalization based on a sample. For example, we have concluded that 95% of graduates can read fairly complex written texts fluently. Suppose that Ann is a graduate. We conclude that Ann is able to read fairly complex written texts fluently. The inference scheme is:

> n% of population S has property P.
> Individual a belongs to population S.
> _____
> a has property P.

This is a statistical syllogism. The argument starts out from a statistical generalization and applies it to a particular case. Note that statistical syllogisms are different from simple predictive inferences, although in both types of reasonings something is concluded about a single individual. Simple predictive inference is a type of inference about the *future*: based on a past sample in which a property is universally present, I conclude that the *next* member of the population will have that property. Statistical syllogism, on the other hand, *is not* a generalization about the *future*: the population we are generalizing about (in our case, graduates) is fully given (we wanted to generalize about *present* graduates) and the particular individual about which something is concluded is part of this population.

9   Statistical reasoning

Let us look at some aspects of the reasoning exemplified above. Implicitly, we have "applied" the statistical value (95%) to a particular case, concluding that there is a 95% probability that Ann can read fairly complex written texts fluently. Then, we have omitted the quantitative specification, because it is very high, and we have simply concluded that Ann is able to read fairly complex written texts fluently. There is no guarantee, of course, that Ann does not fall within the remaining 5%, that is, that she is *not* able to read fairly complex written texts fluently.[1] Second, *the form* of the first sentence, *per se*, does not tells us that the first premise is the conclusion of a (purely) statistical generalization. It could be a truth revealed by God or, more generally, something that has not been concluded based on a statistical survey. That it is the conclusion of a statistical generalization will emerge from a request for justification: "On what grounds do you say that 95% of graduates can read fairly complex written texts?". This question may receive the answer: "It is the conclusion of a statistic". In concrete contexts, a sentence such as "95% of graduates can read fairly complex written texts" is usually asserted because it is the conclusion of a statistical generalization, and therefore the problem we have just outlined is unlikely to emerge. However, it is better to be aware of the fact that the logical form of the sentence in question does not reveal that it is the conclusion of a statistical generalization. "95% of graduates can read fairly complex written texts" is either true or false, just like any sentence that does not contain percentages. How we know it – by a statistical generalization or by a revelation from a hypothetical omniscient source – is another matter.

The closer to 100% or 0% the percentage of the statistical generalization is, the stronger a statistical syllogism is. For example, the argument in the previous paragraphs would be stronger if 99% – and not 95% – of graduates were able to read fairly complex written texts fluently. Notice that if the percentage were 100% or 0% and if the margin of error were zero, then the reasoning would become deductive: if 100% of graduates are able to understand a fairly complex written text, then Ann, being a graduate, will be able to understand such a text. In this case, if the premises are true, so is the conclusion. However, although deductively valid, this argument is not necessarily sound because the truth of the conclusion depends on the truth of the premises and one of the premises is a statistical generalization that might be false. The sample on the basis of which we concluded that 100% of graduates are able to understand a fairly complex written text might have been too small or biased, or the survey might have been poorly conducted and the submitted texts not complex enough. In this case, the chance that the conclusion is true is equal to that of the statistical generalization. By contrast, the chance that the conclusion of

---

1.   Note that the argument that starts out from the premises: "95% of graduates are able to read fairly complex written texts" and "Ann is a graduate", concluding that there is a 95% probability that Ann is able to read fairly complex written texts is a deductive reasoning. It applies something that is true of all individuals in the population (graduates) to a particular case (Ann).

a statistical syllogism starting from a percentage of less than 100% is true is lower than that percentage.

9.4.1 Statistical syllogisms and reference classes

One of the potential problems connected with statistical syllogisms concerns the *reference class* to which the object under examination is ascribed. The reference class is the subset of the population that we consider relevant for our statistical reasoning and for its *application* to particular cases. Returning to the example of the surgical operation discussed in § 9.2.5, John ascribed his elderly sick relative to the reference class of "patients in serious health conditions who had had surgery", and this gave results different from those he would have obtained considering the class of "patients who had had surgery".

An example will clarify how crucial the choice of the reference class is in statistical syllogisms. Suppose a statistical survey shows that only 3% of American environmentalists voted for Trump in the 2020 election. Susan is an environmentalist. We can conclude that Susan *did* not vote for Trump using a statistical syllogism. Our reasoning can be summarized as follows:

> 97% of American environmentalists did not vote for Donald Trump.
> Susan is an American environmentalist.
> ___
> Susan did not vote for Trump.

Let us also assume, though, that Susan is a relative of Trump's, and that a further statistical survey shows that 95% of Trump's relatives voted for him. Applying the same inference scheme, we can conclude that Susan voted for Trump:

> 95% of Trump's relatives voted for Donald Trump.
> Susan is a relative of Trump.
> ___
> Susan voted for Trump.

The conclusions of these two arguments contradict each other. The point, however, is that we have no reason to question the truth of the premises. However, we do not want to run into a contradiction – Susan cannot have voted and not voted for Trump at the same time.

The key for solving this problem lies in the reference class. In particular, if an object can be ascribed to several reference classes that are *relevant* for the statistical reasoning (in our example, the classes of environmentalists and of Trump's relatives), then the *intersection* among all these classes must be considered in statistical syllogism. In our case, we need to take into consideration the set of Trump's relatives who are environmentalists and ascertain what percentage voted for Trump.

9 Statistical reasoning

Suppose you find out that 53% of Trump's relatives who are environmentalists voted for him in the presidential election.[2] How does this affect our reasoning in the case in question? We could reason like this:

> 53% of Trump's relatives who are environmentalists voted for him.
> Susan is both a relative of Trump and an environmentalist.
> ―――――――――――――――――――――――――
> Susan *did not* vote for Trump.

This reasoning is however *bad* because the conclusion is not supported by the statistical basis. After all, the *majority* of Trump's relatives who are environmentalists voted for him. On what grounds can we conclude that Susan behaved otherwise? The inference is clearly *irrational*. Let us take instead:

> 53% of Trump's relatives who are environmentalists voted for him.
> Susan is both a relative of Trump and an environmentalist.
> ―――――――――――――――――――――――――
> Susan voted for Trump.

This inference does not present the same problem: the majority of Trump's relatives who are environmentalists voted for him, and we use this fact to conclude the same thing with regard to Susan. The point, however, is that 53% is a low percentage – it leaves, so to speak, 47 out of 100 possible cases in which Susan *does not* vote for Trump. It is therefore not a sufficiently solid support for the conclusion, and consequently this statistical syllogism is *very weak*, to the point that we would hardly make use of it.

This example shows that when we want to use a statistical syllogism, it is necessary to consider *all* the relevant reference classes and to ascribe the object under examination to the intersection of these classes. An obvious question is which classes are *relevant*. Clearly in a US political election, being an environmentalist, being a relative of the candidate, being for or against the free sale of arms, being for or against the death penalty, etc. are relevant properties. By contrast, wearing a certain shoe size or being a basketball fan are (usually) not. However, there may be many uncertainties about the relevance of a certain property to a statistical syllogism. For example, is gender relevant? Sometimes it might be, sometimes not. We must therefore have a certain pre-understanding of what is relevant to our topic, that is, we must rely on some basic assumptions. Moreover, this is a common characteristic of inductive reasonings, which we have already encountered: usually these reasonings are part of a framework of background assumptions, without which they would be impossible.

---

2. Note that this needs to be ascertained through a further investigation and cannot be inferred from the fact that 3% of environmentalists voted for Trump and 95% of Trump's relatives voted for him. For example, we do not know how many of Trump's relatives are environmentalists. It might well be that the intersection between the two classes is empty.

9.5 Conclusion

Let us recapitulate. We have defined statistical generalizations and we have seen that we can distinguish two types: *purely statistical generalizations*, in which not only the sample but the entire population is fully given, and *generalizations concerning the future*, in which part of the population is not yet given. We have introduced some basic notions such as *sample, population, target property, measured property, accuracy,* and *margin of error*. We have discussed purely statistical generalizations and some of the properties that contribute to their strength or weakness: *significance* and *accuracy* of the measured property against the target property, *size adequacy* and *representativeness* of the sample. We have also considered other elements that contribute to the strength of a statistical generalization: *unbiased* questions and the lack of *hidden variables*. We have discussed the importance of the margin of error value. We have also dealt with generalizations concerning the future, explaining why they are systematically weaker than purely statistical generalizations, and mentioned their relation to scientific laws. We have closed the chapter by discussing an inference scheme that applies statistical generalizations: the *statistical syllogism*. We have explained why the choice of the *reference class* is crucial in this type of inference.

10 | Probability and probability biases

In this chapter we will deal with probability and probabilistic reasoning. First, we will clarify the notion of probability and give some examples of probabilistic statements and problems (§ 10.1). Then, we will discuss three interpretations of probability: the classical interpretation, the frequentist interpretation, and the subjective interpretation (§ 10.2). After that, we will consider the basics of probability theory (§ 10.3), with particular attention to Bayes' theorem and conditional probabilities, and we will exemplify some problems in probabilistic reasoning to which we tend to provide wrong answers (§ 10.4). Finally, we will discuss the related notions of *risk* and *fundamental uncertainty* (§ 10.5).

10.1 Introduction

Take the following problem:

> A cab is involved in an accident at night. There are two taxi companies in the city, one with green cabs and the other with blue cabs. 85% of cabs in the city are green and 15% are blue. An eyewitness claims that the cab involved in the accident was blue. The court is already acquainted with the witness and their reliability in the same kind of cases, and they know that that the witness has correctly identified each of the two colors in question 80% of the time, while failing 20% of the time.
>
> *What is the probability that the cab involved in the accident is blue rather than green, conditional on the fact that the witness, in all good faith, reported that the cab was blue?*

This problem involves the notion of the *probability* of an event, and in order to answer it, we must be able to reason with probabilities. To be more precise, the problem in question requires reasoning with the notion of *conditional probability*. What we are wondering, in fact, is not the sheer probability that the cab is (say) blue, but the probability that the cab is blue *given the fact that* the witness (genuinely) believes that the cab is blue. The events expressed by "The cab is blue" and "Conditional on being identified as blue, the cab is blue" are two different events, and, as we will see, they can have different probabilities. This requires the ability to apply *probability theory*, or at least its most elementary tools. Probability theory is a branch of Mathematics, and is considered by many to be the correct normative theory for our reasoning about *chance*[1] and *uncertainty*. We will discuss the problem again in § 10.4.

As we shall see, reasoning about probabilities is far from trivial: probabilistic reasoning is perhaps the one in which we most systematically tend to deviate from the normative prescriptions – here, those provided by probability theory. At the same time, probabilistic reasoning is of crucial importance, if only because the notion of probability is ubiquitous in our daily considerations, besides being called into question in courtrooms and legal systems, in the valuation of financial risk, in Economics in general, and in Medicine – the latter point being even more evident in a gloomy sort of way, due to the worldwide Covid-19 pandemic that started in 2020.

Take for instance the quantitative conclusion of a recent study:[2]

> Individuals under the age of 60 who have contracted SARS-CoV-2 infection have a 69% chance of not developing clinical symptoms.

Or consider what the *Financial Times* reported in a 2016 article, namely that:

> *Pimco*, the bond fund giant, says there is a 40% chance that Britain will vote to leave the EU in the next referendum. A result that could have a "long-lasting" market impact.

Also in this case we see the evaluation of an event (exit of the UK from the EU) in terms of its probability, and the aim of this evaluation is to predict future market scenarios. As we will see near the end of this chapter, the notion of probability is presupposed by a central notion in financial theories and – more in general – theories in Economics, namely: the notion of *risk*.

In this chapter, we will look at the basic tools for reasoning about probabilities, with a particular focus on reasoning with conditional probabilities. Let us call the whole pack *probabilistic reasoning*. The problem and the statements that we have exemplified above are *probabilistic* problems and statements, as they involve the notion of probability. But *what is* probability?

In a very general way (but also a very satisfactory one, in our opinion), we can say that probability is a *numerical description of how plausible (or likely) the occurrence of a given event is*. In this sense, probability is one of the main tools we have in order to measure chance. However, we can also use probability to measure our *uncertainty* about the occurrence of an event. In principle, there is an important con-

---

1. Here, by the term "chance" we will refer to all those situations in which it is not determined whether a given event occurs (or will occur) or does not (or will not) occur.
2. Piero Poletti, et al., *Probability of symptoms and critical disease after SARS-CoV-2 infection*. arXiv. Preprint published online on 22/6/2020. https://arxiv.org/abs/2006.08471

ceptual difference between the two. *Chance* is an objective feature of the world – some events are not determined: they are just more or less likely to occur. The term "uncertainty", on the other hand, usually refers to our epistemic state: we do not know whether an event does or will occur, but we are more or less confident that it will. However, we will see (§ 10.4) that the term "uncertainty" is also used in another sense in the literature on probability, that is: to denote those aspects of the world that are not determined and regarding which it is not possible to rationally make a probabilistic evaluation. In order not to create confusion, we will use the expression "fundamental uncertainty" when speaking of uncertainty in this sense, when it cannot be measured probabilistically.

Coming back to the difference between measuring chance and measuring uncertainty, it should be noted that those who want to apply probability to (subjective) uncertainty are the proponents of what is termed the *subjective* interpretation of probability, and they argue that our attributions of probability ideally express the *strength* of our *confidence* that an event will occur (see § 10.2.3). This is not the only possible interpretation of probability. In this chapter, we will also briefly introduce two other interpretations. Let us use this preview to make one point clear. The interpretations in question provide three different ways of conceiving probability – if you like, they give us an "intuitive" reading of the notion of probability. All three interpretations, however, agree in saying that, *from a mathematical point of view*, probability behaves exactly as indicated by *the theory of probability* – a branch of Mathematics. In other words, none of the interpretations of probability come up with a formal theory of probability as its product: they all accept the mathematical theory of probability as an explication of how probability works.

Before approaching these three interpretations in detail, one brief remark is worth making. We usually speak of *probability* as something that applies to *events* and measures the likelihood of their occurrence. In this volume we will stick to this usage. In the literature on probability, especially the literature that supports the subjective interpretation, probability is also said to otherwise apply to *propositions* and to measure the likelihood of their being true. This is a totally legitimate use, but we will not adopt it in the following pages.

10.2 Three interpretations of probability

The notion of probability has at least three different conceptual interpretations: that is, there are at least three ways of "reading" this notion and, therefore, probabilistic statements. Let us take a brief look at them.

10.2.1 The classical interpretation

According to the *classical interpretation*, the probability P(*E*) of an event *E* is the ratio between the *number of cases favorable* to *E*, i.e. in which *E* occurs, and the *number of possible cases*, under the *fundamental* assumption that *all possible cases are* equally *possible*. What does this mean? Take a regular, fair die with six faces. What is the probability that 1 will come up on the next roll of the die? There are *six* possible outcomes, and there is no reason to assume that any of them is "more likely" than another. There is, moreover, only *one* outcome every throw: when we roll the die, we just get one result. It is these three points that *justify*[3] our prompt (and correct) answer that the probability of getting 1 on the next throw is 1/6. The classical view only extends the points underlying this example, generalizing them to an overall interpretation of probability. In general, according to this approach, if the possible events are n and the favorable events are nE, the probability of the event *E* occurring will be: P(*E*) = nE/n.

Even when we say that if we flip a coin there is a probability of 1/2 that heads will come out, or that there is the probability of 18/37 that red will come out on roulette, or 1/43,949,268 that a certain winning line will come up at the lotto game, we are assuming that the space of possibilities is *equally distributed* among the *total number of cases*.

A problem with the classical interpretation is that, at least at first sight, it is not a smooth guide to the interpretation of those probabilistic situations – indeed frequent and relevant – in which we cannot assume that the space of possibilities is equally distributed among the totality of cases. In fact, suppose our die is loaded. If this is the case, we cannot know *a priori* what the probability is that the outcome will be, say, 2. Similarly, we cannot know *a priori* what is the probability, in a given population, of developing lung cancer if one smokes a pack of cigarettes a day for 20 years, and we cannot know *a priori* the probability of being hospitalized in intensive care after contracting Covid-19, at the age of 40. In cases like these, we do not have a space of possibility equally distributed among all possible cases. If that were the case, then we would conclude that, in the last two examples considered, the probability of the events in question is 50%. Indeed, for each individual in the population, there are just two possible events: developing or not developing lung cancer (in the first example) and being hospitalized in intensive care or not (in the second example). But it is clearly not the case that the probability of the events in question is 50%.

---

3. The assumption that the die is fair is crucial to the example. If it were loaded, we could not assume that the six possible events are all equally probable, because a loaded die obviously biases the result toward one of the six possible outcomes.

10.2.2 The frequentist interpretation

When we are faced with cases like the two previous ones, we cannot reason under the assumption that the probability space is equally distributed. However, we can ask ourselves – and then observe – how *often* a certain event occurs. For example, we can calculate the number of people who got lung cancer after smoking a pack of cigarettes a day for 20 years as compared with the total number of people who smoke that number of cigarettes per day, or compute the number of 40-year-olds who ended up in intensive care after having contracted Covid-19, as compared with all the people who have had the disease in that age range. We can do the same thing with any kind of definable probabilistic event,[4] including those to which the classical theory of probability is applicable. For example, given a fair die, we can roll it many times. We are also aware that the more rolls we have, the more smokers we analyze, the more people with Covid-19 we consider, the closer our result will be to the probability of the actual event. The frequentist interpretation tells us that this dynamic is implicitly applicable to any kind of assignment of probabilistic values. In particular, this interpretation defines the probability P of an event E as the limit of its (relative) frequency of happening, that is, of occurrences of the event when the number of tests approaches infinity.

$$P(E) = \lim_{n \to \infty} \left(\frac{nE}{n} \right)$$

Thus, in the frequentist tradition, the probability of an event is the *relative frequency* of the set of results of a *probabilistic test* associated with the event (for example, the roll of a die), evaluated on an ideally infinite number of repetitions of the test. The idea is that the more cases we have, the better our approximation will be. If the number of cases available approaches infinity, we arrive at a correct estimate of the probability.

A problem with this interpretation is that it is not applicable if we cannot repeat a test several times. As mentioned, the more times we repeat this, the better our estimate of the probability. This is not always possible, however. Certain events are unique and in situations like these the frequentist approach cannot be used to assign probability to an event. For example, suppose we want to estimate the probability that Spain will win the 2022 football world championship. We cannot apply the frequentist approach because we cannot have the next world cup played a very large number of times: the 2022 football world championship is a unique event. On the other hand, we cannot use the classical approach because we are not facing an equally distributed space of possibility. It does not seem reasonable to attribute equal probability to Spain's victory and Lithuania's victory, for instance. When

---

4. By a "definable probabilistic event" we mean any event to which we wish to and can assign a probabilistic value.

faced with such cases, we end up asking again: what does attributing a probabilistic value to an event mean?

10.2.3 The subjective interpretation

The *subjective interpretation* of probability provides an answer to this question. According to this view, probability measures the *degrees of confidence* or *partial beliefs* that rational agents have on the occurrence or non-occurrence of an event. Let us first explain what we mean by "degrees of confidence". Given any event E, we have greater or lesser confidence in its occurrence.[5] For instance, we can have *more* or *less* "confidence" in the event expressed by the statement "Spain will win the football world championship in 2022". We speak of "degrees of confidence" precisely to refer to the fact that our confidence is "gradable" and it can therefore be stronger or weaker, along a scale. In a somewhat unpolished way, albeit more apposite than it may seem at first glance, we could say that degrees of confidence measure *how much* we believe in an event, that is, the strength of our beliefs. According to the subjective interpretation, the degrees of confidence of a rational agent assign quantitative values to our confidence in an event *according to the rules of probability theory* – we will look at its basics in § 10.3. According to this interpretation, degrees of confidence are therefore attributions of probability values. Or rather, the interpretation conveys the idea that we are rational epistemic agents only if we assign our degrees of confidence towards events by applying the rules of probability theory. Since the attribution of degrees of confidence is more relevant in situations in which it is not certain what happens or will happen, we can say that, for the subjective interpretation, probability is our rational tool for measuring uncertainty.

Let us come back to the above example. If we say that Spain has a 15% chance of winning the next world cup while Lithuania's is only 0.05%, this reflects our degree of confidence in the realization of these events and indicates that our confidence as regards Spain's winning the cup is far stronger than our confidence regarding Lithuania's winning the cup. Generally, sports betting winnings are computed on this basis. The same concept underlies the probabilities assigned to weather forecasts or forecasts on economic trends or the performance of a company next year, on which, for instance, the performance of the stock exchange, or of a stock, depends. According to the subjective interpretation, in other words, probability is precisely this: each of our probabilistic attributions is actually the quantitative expression of how much rational confidence we have in the occurrence of an event.

To say that probability is *subjective*, because confidence in events obviously is such, is not the same as saying that it is *arbitrarily assigned*. Our degree of confi-

---

5. Of course, this is relevant only when we do not know whether the event will occur or not.

dence in the occurrence of an event very often has rational grounds: I have a greater degree of confidence in Spain's victory than Lithuania's victory because I know that Spain has very strong players, and because I know that Lithuania will not be able to line up players of that level. Under these conditions, I am *rationally justified* in assigning a greater degree of confidence to Spain's victory than to Lithuania's. It is because the satellite photos show a strong typhoon that is reaching the area where I live that I am rationally justified in having a great degree of confidence in the fact that it will rain tomorrow. It is because I know that the pandemic has forced companies to adopt smartworking in a massive way that I am rationally justified in having a great degree of confidence that certain tech stocks will grow in the equity markets. The subjective interpretation does not encounter difficulties even in the face of the paradigmatic cases that motivate the classical view: where we can apply the assumption that the probability space is equally distributed, the subjective interpretation tells us that rational agents will have the same degree of confidence regarding each of the events defined in the probability space. For instance, under the subjective interpretation, when facing the roll of a fair die, agents are rational only if they have equal confidence in every possible outcome, and in particular if they attribute 1/6 to each of them.

Nevertheless, the subjective interpretation of probability is "really" *subjective*, in the sense that probability assignments vary from agent to agent and from time to time, depending on the information available to agents. For instance, if a meteorologist acquires more detailed satellite images, better and more defined isobaric maps at different altitudes, and better maps of the winds, his confidence in a certain forecast regarding the weather in three days will likely change, and consequently the probability assigned to a certain forecast will also change. On the other hand, meteorologists who have more information and maps available than other meteorologists will give a certain forecast a different probability from that given by their colleagues who have less.

That being said, an important problem that the subjective interpretation has to tackle is the "assignment of *prior* probabilities".[6] It is not always clear on what ground we assign a given value rather than another to the degree of confidence we have in a given event. Let us see what this means. In the previous example we attributed the value of 15% to our degree of confidence in Spain's victory, and 0.05% to the degree of confidence in Lithuania's victory, and we explained why that difference may be rational, at least in principle. But on what grounds is the probabilistic value we attribute to Spain's victory 15% and not 10%? And on what grounds is this value, in the case of Lithuania, 0.05% instead of 0.1%? In this case too (and for

---

6. As for the difference between *prior* probability and *posterior* probability, we refer the reader to § 10.3.4. Here, suffice it to say that the *prior* probability of an event is the unconditional probability of the event, that is, a probability that does not depend on the occurrence or non-occurrence of another event.

the very same reasons as above) we could be rationally justified in attributing these values to the two events. Why one pair of values rather than the other? It is hard to answer, because there is no objective aspect of the world that justifies one over the other. In short, it seems that, although the "subjective" view of probability does not collapse into a fully arbitrary view of probability, there is, to some extent, a somewhat arbitrary component in the attribution of degrees of confidence in an event.

In any case, it is very important to understand "where" this arbitrariness lies, that is (at least in some cases) in the choice of given quantitative values. There is no arbitrariness, however, when it comes to what the epistemic agent "ought to do" with these values: they must conform to the rules of probability theory. Ideally, once the values for the basic events are set, an agent operates on them in accordance with those rules. This is because the subjective interpretation focuses on the *internal coherence* of our degrees of confidence rather than on their assignment based on objective aspects of the world. This explains why the subjective interpretation is a "coherentist" conception of probability.

10.3 Basic aspects of probability theory

Let us now turn to some basic aspects of probability theory. As we have already said at § 10.1, the way probabilities are to be computed is independent from the different possible interpretations of probability. In short, we have at least three conceptual interpretations of probability, but only one (mathematical) theory of probability. Here we will see how to compute the probabilities, or rather, given the probabilities assigned to the elements of a set of events (whatever the procedure used to assign them), we will see how we should derive the probability of certain combinations of the events of the set. Here, "combination" is to be understood in a very broad sense. For instance, the fact that two events occur together, or that either of them occurs, or that one occurs as long as the other has occurred, are all combinations of events in the terminology that we are using here.

Probability theory is one of the most recent branches of Mathematics. Indeed, before the seventeenth century, no treatment of the notion of "probability" in mathematical terms had ever been attempted, and the notions of "probable" or "likely" were vague and did not allow for any rigorous reasoning about chance or, even less, for measuring chance. In the 17th century, the foundations were laid for a mathematical treatment of this notion due, to a good extent, to the fundamental contributions of Blaise Pascal, Pierre de Fermat and Christiaan Huygens. Indeed, it is usually said that the history of probability theory really begins in that century. It is important to notice two things here. First, before the 17th century we did not really have a way to think about "chance" events, those events whose occurrence is not determined. Second, probability theory has provided a rigorous definition for a concept that was vague before the theory was built, and it has thus somewhat redefined the notion of probability. Most importantly, probability must satisfy the rules that the corresponding mathematical theory has defined for it. If we are reasoning with "probability", but not in accordance with those rules, then we are not *really* reasoning with probability, because those rules *define* probability – what we are doing is, rather, using some concept that has some kind of family resemblance to probability. It is also worth noting that human beings are, on average, "fragile" probabilistic reasoners, more fragile than they are in other areas of reasoning. That is, we tend to be mistaken and are easily misled when we think about the probability of events. Perhaps this is due to the very fact that the theory of probability has only recently provided precise rules for our reasoning as regards chance. In short: apparently, we tend to think about chance in terms that are either prior to the theory of probability, or not entirely consistent with its notions.

We shall begin by introducing some basic terms. In order to assign probabilities, we must first define the space of events, which we will call *probabilistic space*[7] in the rest of this chapter. This is the space of events that is defined by the probabilistic problem we are considering. Intuitively, it consists of the set of all possible results of a probabilistic test. We denote it with the symbol Ω. For instance, the flip of a coin determines an event space that includes *two* basic events ("heads" and "tails"), and all the events that can be defined from them using a series of operations that we will look at shortly. Probability theory defines *events* as sets of possible outcomes of a probabilistic test. Ideally, we can start with the basic or "smallest" test results. In the case of a coin toss, these are "heads" and "tails". In the case of a roll of a die, the smallest results are the six possible outcomes, namely: "1 comes out", "2 comes out", ..., "6 comes out". Moreover, we can define other events starting with these "smallest" ones, and do things with these "derived" events – for instance, we can bet on them. In addition to the six events illustrated in the case of the roll of a die, we can also consider the event "1 *or* 2 comes out", which in probability theory consists in the *union* of two of the basic events, or the event "2 *does not* come out" (definable, in this case, as the union event of "1 comes out", "3 comes out", "4 comes out 4 ", ...), which in probability theory is considered as the *complement* event of "2 comes out". In general, we can use all the set-theoretical operations to build events out of other events. Furthermore, *each* application of set-theoretical operations to events gives rise to other events. Again, in the case of a single die roll, for instance, we can consider (say) the impossible event "2 comes out and 2 does not come out".

Let us now introduce some notation that is useful for reasoning with probability, and comment on it.

a. Ω will denote the space of events – or "total event";
b. Z will denote the so-called "empty event" – or "impossible event";

---

7. Another name for the probabilistic space is "sample space".

c. P(E) will denote the probability of event E;
d. ¬E will denote the event that is the complement of E. In other words, ¬E means that E does not occur (complement event);
e. E ∩ E' means that both event E and event E' occur (intersection of events);
f. E ∪ E' means that at least one of the events – E and E' – occurs i.e. that either E or E' occurs (union of events).
g. E'|E indicates that E' occurs given that E has occurred.

If an event has probability 1, then it is *necessary* for it to occur (or, in the subjective interpretation, it is *certain* that it does); if it has probability 0, then it is *impossible* for it to occur (or, in the subjective interpretation, it is *certain* that it *does not* occur). For example, the event "tomorrow it rains or it does not rain" has probability 1, while the event "tomorrow it rains and it does not rain (at the same time in the same area)" has probability 0, because it is a *contradictory* or *impossible* event. Events with probability between 0 and 1 are called *contingent*. Notice that P(Ω) = 1 is valid in probability theory, and it is in fact assumed as an axiom of the theory. The axiom can be read as follows: "it is necessary (or certain) that something happens". From this axiom and from the rule for negation that we will see in § 10.3.3, it immediately follows that P(Z) = 0 – notice that the fact that Z = Ω – Ω plays a role in this.

We can then go on and compute the probability of some combination of events given the probabilities of single events. In what follows, the concepts of (*a pair of*) *independent events* and (*a pair of*) *dependent events* will play a crucial role. An informal definition of these two concepts goes as follows: E and E' are two *independent* events if and only if the probability of one occurring is not affected by the probability of the other occurring. Otherwise, they are two *dependent* events.

Examples of independent events are "heads on the first coin" and "heads on the second coin" when tossing two different coins. The probability that the first coin comes up heads is not affected by the probability that the second coin comes up heads (or tails). Instead, here are two pairs of dependent events: "tomorrow there will be a flood" and "tomorrow it will rain heavily"; and "drawing a king on the first draw" and "drawing a king on the second draw" from a full and regular deck of French-suited cards. The drawing of the first card, whether it is a king or not, will affect the probability that a king will come out on the second draw, and the fact that it rains abundantly tomorrow influences the fact that tomorrow there will be a flood.

In the case of two independent events E and E', the fact that the probability of one does not affect that of the other is formally expressed through the fact that P(E|E') = P(E). More precisely, in the case of two independent events we have P(E|E') = P(E) = P(E|¬E'). Indeed, if E' has no influence on E, then the probability of E remains the same whether E' occurs or not. By contrast, if E and E' are dependent events, we have P(E|E') ≠ P(E).

10.3.1 Rules for intersection

The probability that both E and E' occur (or the *probability* of the *intersection* of E and E') is given by the following formula:

$$P(E \cap E') = P(E) \times P(E'|E) \text{ that is, } P(E'|E) = \frac{P(E \cap E')}{P(E)} \text{ with } P(E) \neq 0$$

The probability that both events occur is given by the probability that the first occurs *times* the probability that the second occurs given the first. Since in the case of independent events we have $P(E|E') = P(E)$, this allows us to give a simplified definition for the specific case in which E and E' are independent:

$$P(E \cap E') = P(E) \times P(E')$$

In general, if $E_1 \ldots E_k$ are independent events, then:

$$P(E_1 \cap E_2 \cap E_3, \ldots, E_k) = P(E_1) \times P(E_2) \times P(E_3) \times \ldots \times P(E_k)$$

For instance, the probability of getting two heads by tossing two fair coins is:[8]

$$P(T_1 \cap T_2) = P(T_1) \times P(T_2) = 1/2 \times 1/2 = 1/4$$

This is immediately evident given the assumption of equiprobability of the four possible results, which are $H_1 \cap H_2, H_1 \cap T_2, T_1 \cap H_2, T_1 \cap T_2$.

By contrast, if the events involved are dependent events, we must use the general version of the formula. For instance, assume that K_1 and K_2 denote the event of getting a king on the first draw then another king on the second from a 52-card deck *when the first card is not placed back in the deck after being taken out of it*.

$P(K_1) = 4/52$ [there are 4 kings in a deck]
$P(K_2|K_1) = 3/51$ [51 cards are left, and 3 kings]
$P(K_1 \cap K_2) = (4/52) \times (3/51) = 12/2652 = 0{,}004$

The following is a very similar example. Suppose that there are 18 balls in a box, 5 of which are red, 6 are blue and 7 are yellow. What is the probability of taking two red balls in a row, without placing the first ball we took out back in the box?

$$P(R1 \cap R2) = (5/18) \times (4/17) = 20/306 = 10/153$$

What is the probability that the first ball we take out is blue and the second is yellow? It is the following:

$$P(B \cap Y) = (6/18) \times (7/17) = 42/306 = 7/51$$

---

8. In these examples, we let H_1 and H_2 denote the two events of type "heads", and T_1 and T_2 the two events of type "tails" with the indexing referring to the corresponding toss.

10.3.2 Rules for union

The probability that either E or E' occurs (that is, the *probability* of the *union* of E and E') is determined by the following formula:

$$P(E \cup E') = P(E) + P(E') - P(E \cap E')$$

That is, the probability that either E or E' occurs is given by the sum of the probabilities of the two events *minus* that of their intersection. We will comment on the formula below, and we will show its adequacy. For now, let us see a special case of it, which presupposes the notion of "mutually exclusive events". Two events E and E' are *mutually exclusive* if and only if they cannot occur together, that is if $E \cap E' = Z$. Since $P(E \cap E') = 0$ if $E \cap E' = Z$, clearly, if two events E and E' are mutually exclusive, then $P(E \cap E') = 0$. In this special case, the general formula can be exemplified as follows:

$$P(E \cup E') = P(E) + P(E')$$

For instance, the probability of drawing a king (K) or an ace (A) of any suit whatever from a 52-card deck on just one extraction is:

$$P(K \cup A) = (4/52) + (4/52) = 8/52 = 0.15$$

Indeed, since we have 4 kings and 4 aces in the deck, the probability of getting a king is 4/52 and that probability is the same as the probability of getting an ace. These are two mutually exclusive events, since we get either a king or an ace (or none of them) with a single extraction, but we cannot get both a king and an ace by it.

Similarly, if in a box you have 18 balls of which 5 are red, 6 are blue and 7 are yellow, then the probability of getting a blue ball or a yellow ball is:

$$P(B \cup Y) = (6/18) + (7/18) = 13/18$$

If two or more events exhaust the probability space, then their union has value 1. For instance, what is the probability that we get either heads or tails by tossing one coin?

$$P(H \cup T) = 1/2 + 1/2 = 1$$

Therefore, it is necessary that we get either heads or tails.

Let us now examine the case in which E and E' are not mutually exclusive – that is, they can occur together. In this case the notation is $E \cap E' \neq Z$. In order to compute the probability that either of two non-mutually exclusive events E and E' occurs, we must use the general formula that we have introduced above:

$$P(E \cup E') = P(E) + P(E') - P(E \cap E')$$

Now we will see why this general formula is adequate. In order to prove its correctness, consider that:

$$P(E \cup E') = P(E') + P(E \cap \neg E') \text{ [recall that } E \cup E' = E' \cup (E \cap \neg E')]$$

Also:

$$P(E) = P(E \cap E') + P(E \cap \neg E') \text{ [recall that } E = (E \cap E') \cup (E \cap \neg E')]$$

Subtracting member to member we then get:

$$P(E \cup E') - P(E) = P(E') + P(E \cap \neg E') - [P(E \cap E') + P(E \cap \neg E')]$$

Notice now that $P(E \cap E') = P(E' \cap E)$; by simplifying the formula we get the formula that we wished to prove. The formula we have introduced above:

$$P(E \cup E') = P(E) + P(E') - P(E \cap E')$$

is actually valid for every kind of independent event, both for mutually and non-mutually exclusive events. When E and E' are mutually exclusive, then $P(E \cap E') = 0$, since it is impossible that the two events occur together and then, as we have seen, we can simplify the formula.

Let us give some examples of the probability of the union of two non-mutually exclusive events. The probability of getting one "heads" if we toss two coins at the same time is:

$$P(H_1 \cup H_2) = 1/2 + 1/2 - (1/2 \times 1/2) = 1 - 1/4 = 3/4$$

The probability of getting at least one number 6 when we roll a pair of dice is:

$$P(S_1 \cup S_2) = 1/6 + 1/6 - (1/6 \times 1/6) = 2/6 - 1/36 = 11/36 \approx 0.30$$

10.3.3 Rule for complement event

The rule for a complement event determines the probability that an event *does not* occur on the grounds of the probability that the same event occurs. This would be:

$$P(\neg E) = 1 - P(E)$$

The formula is easily proven. Indeed, it is necessary that an event either occurs or does not occur. Hence:

$$P(E \cup \neg E) = 1$$

But E and $\neg E$ are by definition mutually exclusive. Hence, $P(E \cup \neg E) = P(E) + P(\neg E)$, from which we get:

$$P(E) + P(\neg E) = 1$$

from this, the rule for complement event follows.

Let us try to apply the rule by computing the probability of getting one heads by tossing a pair of fair coins again. This event occurs when it is not the case that the outcome of the toss is tails for both coins. But we know that the probability of that outcome is 1/4. Therefore, the probability of getting at least one heads is 1 − 1/4, that is 3/4.

There is one further reason why the rule for the complement event is important. We have seen what formula determines the probability of at least one of two independent events occurring. How do we determine this very probability, when the two events are dependent? For instance, how do we determine the probability of getting at least one black ball from a box containing two black balls and three white ones by having two extractions (without putting the first ball back into the box)? Take the following method: we compute the probability of *not* getting any black ball *at all* first. This equates with the probability of extracting two white balls:

$$P(W_1 \cap W_2) = P(W_1) \times P(W_2|W_1) = 3/5 \times 2/4 = 6/20$$

Then, we apply the rule for the complement event:

$$P(B_1 \cup B_2) = 1 - P(\neg(B_1 \cup B_2)) = 1 - P(W_1 \cap W_2) = 1 - 6/20 = 14/20 = 7/10$$

Notice that $P(\neg(B_1 \cup B_2))$ denotes the probability that no black ball comes out from either the first or the second extraction, which equates with $P(W_1 \cap W_2)$, that is, the probability of the two balls that we get both being white.

10.3.4 Bayes' theorem

Bayes' theorem is a fundamental theorem of probability theory, as it gives us the recipe to update a probabilistic value once new information has been acquired. In this sense, it helps us grasp a "dynamic" aspect of probabilities – more precisely, it helps us understand how, exactly, probabilities *change* when certain *events* that are defined in the probabilistic space occur. Here we use the phrases "new information" and "occurrence of an event" interchangeably, due to an idealization – taken from probability theory – which presumes us to be "perfect" observers of all events. From this idealization, it follows that an event defined in a probabilistic space occurs if and only if we know that it occurs. Thanks to Bayes' theorem, which we owe to Thomas Bayes (1702-1761), we have a mathematical formula that provides us with a precise way of "correcting" a probabilistic value in the light of new information. More specifically, if there is a preliminary (*prior*) estimate of the probability of an event, and if additional information is then acquired, Bayes' theorem helps us determine the *posterior* probability of the occurrence of that event. Just to get an idea: what we call *prior* probability is the attribution of probabilistic value to the events of a space "before" any event occurs. It is, for example, the probability we assign to the "2 *does not* come out" event before the roll of a fair die. In this case, the probability of the event is 5/6. A *posterior* probability, on the other hand, is the attribution of probabilistic value to events "after" one of them has occurred. After 2 turns out to be the outcome of the roll of the die in question, the probability of the event "2 *does not* come out" (with respect to the throw just given) "becomes" 0. Obviously, it "would become" 1 if any number other than 2 came out. Similarly, the 4/52 probability of getting a king from a regular, freshly opened deck of French-suited cards is a *prior* probability, while the 3/51 probability of getting a king from a deck of French-suited cards *from which one king has already been removed* is a *posterior* probability. Notice that the posterior probability of an event E is always relative to an event E' which we assume has occurred. We will speak just of "*posterior* probability" of an event E, without further qualification, when we have already introduced the event that we assume occurred, and the context will thus make it clear to which event the *posterior* probability of E is related. Given what we have just said, maybe you have already guessed that *posterior* probability is conceivable in terms of *conditional* probability.[9] Therefore, it is not surprising that Bayes' theorem can be read more generally as a formula which captures the relationships between different conditional probabilities, as we will see shortly.

Like all the tools of probability theory we deal with in this section, Bayes' theorem is valid in the context of all the interpretations of probability – the subjective interpretation, as well as the frequentist one and the classical one. The theorem evaluates the updated or posterior probability of two *mutually exclusive* but *jointly exhaustive events*. Two events are jointly exhaustive when they exhaust the space of possibilities: either one occurs or the other does, but they cannot occur together. For instance, E and $\neg E$ are mutually exclusive and jointly exhaustive events. The theorem can be extended to a number n of mutually exclusive but jointly exhaustive events, but here we confine ourselves to just two events for simplicity. The theorem is particularly useful for updating the probability of a certain event when we get more information about that event.

Suppose that E and E' are two mutually exclusive but jointly exhaustive events, and that E'' is a third event, which constitutes the new information that has occurred. Then, the theorem has this form:

---

9. Actually, the *prior* probability of an event E is also relative to another event, that is, to the "total event" which, as we have seen, can be read as "something occurs". Of course, this being given, also *prior* probability can be represented in terms of conditional probability. Specifically, $P(E) = P(E|\Omega)$. The formula is secured by the fact that $P(\Omega) = 1$.

$$P(E|E'') = \frac{P(E) \times P(E''|E)}{[P(E) \times P(E''|E)] + [P(E') \times P(E''|E')]}$$

We can think of P(E) as the *prior* probability of E and of P(E|E'') as the conditional probability of E given E''. Since this is the event that we assumed to occur, we can take *P(E|E'')* to be the *posterior* probability of E. Notice that, since E and E' are mutually exclusive but jointly exhaustive, $E' = \neg E$, namely the occurrence of E' *is* equivalent to the *non-occurrence* of E. Hence, the above statement of the theorem is equivalent to the following formula:

$$P(E|E'') = \frac{P(E) \times P(E''|E)}{[P(E) \times P(E''|E)] + [P(\neg E) \times P(E''|\neg E)]}$$

Now let us see how the theorem can be proved. Remember the general formula for calculating the probability of the intersection between two events:

$$P(E \cap E'') = P(E) \times P(E''|E)$$

Applying this very formula by reversing the order in the intersection, we get:

$$P(E'' \cap E) = P(E'') \times P(E|E'')$$

Since $E \cap E'' = E'' \cap E$, these two formulas are equivalent, so:

$$P(E'') \times P(E|E'') = P(E) \times P(E''|E)$$

By dividing both sides of the equivalence by $P(E'')$ we obtain:

$$P(E|E'') = \frac{P(E) \times P(E''|E)}{P(E'')}$$

We already know from the proof of the rule for union in § 10.3.2 that:

$$P(E'') = P(E'' \cap E) + P(E'' \cap \neg E)$$

That is, the probability that an event occurs is given by the sum of the probability that that event occurs together with another event with the probability that the event occurs without the other event occurring. Hence:

$$P(E|E'') = \frac{P(E) \times P(E''|E)}{P(E'' \cap E) + P(E'' \cap \neg E)}$$

We also know from the general rule for the probability of the intersection of two events that:

$$P(E'' \cap E) = P(E) \times P(E''|E)$$
$$P(E'' \cap \neg E) = P(\neg E) \times P(E''|\neg E)$$

Substituting in the denominator of the fraction, we get

$$P(E|E'') = \frac{P(E) \times P(E''|E)}{[P(E) \times P(E''|E)] + [P(\neg E) \times P(E''|\neg E)]}$$

and form this, we conclude:

$$P(E|E'') = \frac{P(E) \times P(E''|E)}{[P(E) \times P(E''|E)] + [P(E') \times P(E''|E')]}$$

that is, the formula stated by the theorem. Notice that the theorem also gives us the ratios between the conditional probability of E given E'', on the one hand, and the probability of E, of E'' given E, of E', and of E'' given E' on the other. The importance of Bayes' theorem lies in the fact that, given two mutually exclusive and jointly exhaustive events E and E', it gives us a way to compute the conditional probability of E given any third event E'', providing that we have the value of the prior probability of E and that of the conditional probability of E'' given E. As for event E', we can always take $\neg E$ in its place: since E and E' are mutually exclusive and jointly exhaustive, obviously, $E' = \neg E$. Applying this to situations in which it is assumed that E'' has occurred, Bayes' theorem provides us with a procedure for determining the posterior probability of E, provided – again – that we know the value of the *prior* probability of E and that of the conditional probability of E'' given E. Under this reading, the theorem also reveals to us that the posterior probability of an event E is completely pre-encoded into the conditional probability of E, given an event E'' that then happens to occur. The change of probability values following the contraction of the probabilistic space occurs along lines already codified before the events occur.

Bayes' theorem has such a wide range of applications that it is impossible to list them all. A classic and fairly immediate application is updating the probability that a subject has a certain disease after having tested positive on a medical examination. For example, suppose a disease affects 2% of people. This implies that any person has a 2% chance of having the disease. Suppose a person undergoes a medical examination aimed at diagnosing the disease. We also assume that the test has an 80% reliability. Obviously, that increases the probability that that person has the disease. But how much does that increase such a probability, exactly? We can use Bayes' theorem to establish this. E denotes the fact that the subject has the disease, E' denotes a positive outcome of the test. $P(E)$ is the probability that the subject has the disease prior to the test (0.02). $P(E|E')$ is the probability that the subject has the disease given a positive outcome of the test. $P(E'|E)$ denotes the likelihood that the test gives a positive outcome provided that the subject has the disease (0.8 because the test is 80% reliable). $P(E'|\neg E)$ indicates the likelihood of being positive to the test despite not having the disease (which is 0.2). Hence:

$$P(E|E') = \frac{0.02 \times 0.8}{(0.02 \times 0.8) + (0.98 \times 0.2)} = \frac{0.016}{0.016 + 0.196} = \frac{0.016}{0.212} \approx 0.075$$

Therefore, a patient that is positive on the test has about a 7.5% chance of having the disease. It is certainly higher than the initial 2%, but perhaps less than one would expect. Obviously a more reliable test would imply a greater increase in the probability of having the disease.

It is worth mentioning some "historical" cases in which Bayes' theorem has been applied, and these are all from outside Medicine. For example, it was used in the search for Air France flight 447 from Rio de Janeiro to Paris, which disappeared in June 2009. The initial two-year searches were unsuccessful but, six days after applying Bayes' theorem, the airplane was found 4,000 meters deep in the Atlantic Ocean. Various types of data were used in the application of the theorem, including the exact path followed, wind speed, ocean currents, etc. Each of these pieces of evidence changed the probability of finding the plane in a certain area, allowing searchers to identify the area where the plane had most likely crashed. Bayes' theorem was also used to locate U-boats during World War II and in the search for an H-bomb following an accident involving a B-52 in Spain. Obviously, use of the theorem is not always successful, as is to be expected in inductive reasoning. For example, it was unsuccessful when it was employed in the search for Malaysia Airlines Flight 370 from Kuala Lumpur to Beijing, which disappeared in March 2014, somewhere in the southern Indian Ocean.

Bayes' theorem is also involved in many applications of probabilistic reasoning to everyday life, as we will see below.

10.4 Two probabilistic reasoning problems, and some fallacies

In this section we will show some probabilistic reasoning problems, which help us understand, among other things, how easily we are misled when it comes to reasoning with probabilities. Kahneman and Tversky, who we have already mentioned in Chapters 1 and 6, have studied the fallacies we tend to make in probabilistic reasoning in great detail.

Here we report one of the many fallacies of probabilistic reasoning they studied. The fallacy is known as the "conjunction fallacy" or "Linda problem". It shows the tendency to make mistakes in thinking about the probability of the intersection of two events. The first name of the fallacy is due to the analogies between conjunction (sentential connective) and intersection (set-theoretical operation). The fallacy is interesting precisely because it violates a very elementary rule which is very easy to understand, and this in turn shows just *how* fragile we are as probabilistic reasoners. Kahneman and Tversky presented the following story to some students:

> Linda is a thirty-year-old single woman. She is outspoken and very bright, and she graduated in philosophy. As a student, she was much involved in the public issues of discrimination and social justice. She also took part in anti-nuclear rallies.

After presenting the story, Kahneman and Tversky asked the students: Which of the following two options is more probable?

1. Linda is a bank employee.
2. Linda is a bank employee and she is involved in the feminist movement.

Most students took 2 to be the right answer, although probabilistic reasoning does not justify this answer – by contrast, it justifies only 1. Indeed, no matter how strongly we feel that Linda must be involved in the feminist movement, given the description, the probability of the intersection of two events (and 2 just offers that) will always be lower than (or equal to) that of the individual events involved, as is clearly proved by the rule for intersection (§ 10.3.1). Therefore, when answering the question, most of the students choose the wrong one, and they violate one of the most elementary rules of probabilistic reasoning. Why? Because *de facto* we tend to reason according to criteria of "sample representativeness". Linda's description makes her extremely representative, at least in our imagination, of the profile of a person who is involved in the feminist movement, and we therefore tend to think that of the two options, the correct one is the one that actually incorporates the consequences of this assumption of representativeness. In our case, this is option 2, not option 1. However, this does not mean that the answer to the question is correct, as the question is about probabilities and, by choosing 2, we violate a basic rule of probability theory.

Despite this, the experiment does not necessarily show that we are "hopelessly fragile" when we reason with probabilities. Indeed, it may be that we implicitly read option 1 as (1a) "Linda is a bank employee *and she is* not *involved in the feminist movement*", adding the second conjunct precisely because we think this fits the context, as the other option gives us "Linda is a bank employee and is involved in the feminist movement". This does not mean that the answer given by the majority of the students is justifiable, because the answers are in any case 1 and 2, not 1a and 2. This consideration, moreover, does not contradict the hypothesis of Kahneman and Tversky according to which this wrong answer is due to the influence of considerations on Linda's representativeness. However, it tells us that, implicitly, students may have reacted by thinking which of 1a and 2 was more probable. If so, ideally they would not have attributed a greater probability to an intersection of events than the probability attributed to one of the individual events.

Let us now look at the two probabilistic reasoning problems involving conditional probabilities.

10.4.1 The cab problem

Let us return to the problem at the beginning of this chapter:

> A cab is involved in an accident at night. There are two taxi companies in the city, one with green cabs and one with blue cabs. 85% of cabs in the city are green and 15% are blue. An eyewitness claims that the cab involved in the accident was blue. The court has already experienced the witness's reliability under the same circumstances as those of the incident, and it knows that that the witness identified each of the two colors *correctly* 80% of the time, while failing 20% of the time.
>
> *What is the probability that the cab involved in the accident is blue rather than green, conditional on the fact that the witness, in good faith, reported that the cab was blue?*

One natural thought, when facing this problem, is that "the witness is likely right" (after all, he – or she – is right 80% of the time!), and we can reasonably interpret this statement as "the probability that the cab has the color the witness claims it has" is greater than 0.5 (or 50%, which obviously is the same). We set the following events:

- B = the cab is blue;
- Y = the cab is yellow;
- CB = the witness (faithfully) asserts that the cab is blue;
- CY = the witness (faithfully) asserts that the cab is yellow.

The problem requires establishing a conditional probability, and we have all the necessary conceptual tools to solve it by means of Bayes' theorem. Indeed, the theorem helps us update the probability that the cab was blue given that the witness says so, that is $P(B|CB)$. We can apply it because we have a value for $P(B)$, i.e. 0.15, a value for $P(Y)$, i.e. 0.85, a value for $P(CB|B)$, i.e. 0.8, and a value for $P(CB|Y)$, i.e. 0.2:

$$P(B|CB) = \frac{P(B) \times P(CB|B)}{P(B) \times P(CB|B) + P(Y) \times P(CB|Y)} = \frac{0.15 \times 0.8}{(0.15 \times 0.8) + (0.85 \times 0.2)} \approx 0.41$$

The fact that we tend to overestimate the probability that the cab is of the color recognized by the witness may be due to one or more elements that are "delicate" to us. First, we actually tend not to distinguish between the probability of E given E', and the probability of E' given E. Or rather, we tend to switch them when we are asked to assume which one is relevant in our reasoning, as will be evident in what is called the "prosecutor's fallacy". Second, the formula of Bayes' theorem is cognitively complex enough for us to induce us to go with an intuitive (but systematically error-prone) estimate of the conditional probability that is of interest for us, instead of applying the proper probabilistic computation.

10.4.2 The prosecutor's fallacy

A further example of how hard it can turn out to be for us real cognitive agents to apply Bayes' theorem is provided by the "prosecutor's fallacy". The name stems from the fact that this erroneous reasoning can show up in court, where it leads to overestimate the probability that an accused is guilty.[10] For example, suppose a crime has been committed in a city where 5 million people live. Jones is charged with the crime, a DNA sample of Jones is taken and it is verified that it matches a biological trace found at the crime scene. The DNA test only fails in one in 10,000 cases. Thus, the prosecutor claims that there is only a 1/10,000 chance that Jones is innocent. Since this is a very small probability, the prosecutor argues that we have sufficient evidence that Jones is the culprit.

This reasoning sounds convincing, but it is wrong. Let us see why. First, we will give the case a rigorous formulation. We suppose that:

- G = Jones is guilty;
- E = the DNA test gave a specific result (namely: the one which we have mentioned above).

In the traditional exemplification of the fallacy, corresponding to the procedural scenario described above, the prosecutor bases his indictment on $P(E|G)$, that is, on the probability that the accused turns out to be positive on the DNA test, *given* that the accused is guilty. $P(E|G)$, as we said, is 9999/10000 = 0.9999. A probability close to 100%, then! The prosecutor concludes that Jones is (*almost certainly*) guilty.

Now let us see why this conclusion is unjustified. What we actually want to know is the probability that Jones is guilty *given the test result*. However, what the prosecutor does is to look for the likelihood that the test will give a certain result given that Jones is guilty. But this is not what we want to know, obviously, if only because the fallacy committed by the prosecutor starts out from the *assumption* that Jones is actually guilty, and "conditionalizes" on it, that is, the prosecutor computes the value of $P(E|G)$. But we do not know whether Jones is guilty or not. Instead, we want to know what the probability is that Jones is guilty given an event that we know (that is, the response given by the DNA test and what it says about the biological sample). In other words, the prosecutor should have reasoned out of

---

10. The fallacy has actually been used in court. One famous case in which this happened is the Sally Clark case. Sally Clark was a British woman. In 1998 she was charged with the murder of her two babies (8 and 11 weeks old at time of death). The woman claimed that they had both died by sudden infant death syndrome (SIDS), but the prosecutor claimed that there was a probability of 1/73,000,000 that two deaths by SIDS occurred in the same family. Sally Clark was convicted but, after a few years, the Royal Statistical Society showed that this was due to a mistaken assessment of probability. In light of this, Sally Clark was freed in 2003. However, what happended got her a serious psychological damage and an addiction to alcohol that led her to death in 2007.

$P(G|E)$, which is the probability that Jones is guilty given the result of the DNA test. If we do so, we come to a completely *different* conclusion. Indeed, $P(G|E)$ can be computed using Bayes' theorem. With 5 million people living in the city, the chance that Jones is guilty *before* the DNA test is 1 in 5 million. After the test, this probability clearly increases. But how much does it increase? $P(G)$, before the test result, is 0.0000002. $P(E|G) = 0.9999$, while $P(E|\neg G)$ – the probability that, despite not being guilty, Jones is positive on the DNA test – is equal to $1/10000 = 0.0001$. Applying Bayes' theorem:

$$P(G|E) = \frac{P(G) \times P(E|G)}{[P(G) \times P(E|G)] + P(\neg G) \times P(E|\neg G)} \approx 0.1996$$

Despite the positive test, there is only about a 20% chance that the accused is guilty. This is certainly a great enough probability to proceed with further investigation by the police about Jones, but it is certainly not great enough to convict him. This tells us that the evidence we need in court cannot be limited *only* to the DNA examination, but must always be accompanied by other clues. For example, if the accused knew the victim, this increases his chances of guilt. If he was on bad terms with the victim, the probability increases further. And so on.

There are two important things that we can learn from the "prosecutor's fallacy". The first is that, as we have already mentioned, in our reasoning, we tend to "switch" the events *on which* we have to "conditionalize" (in our case, the response of the DNA test) and events that we have to "conditionalize" (in our case, Jones's guilt).[11] This highlights once again our great fragility in probabilistic reasoning. Second, the example helps us understand the importance of correct probabilistic reasoning. Failure to reason correctly can lead to someone being sent to jail without justification.

10.5 Risk and fundamental uncertainty

Here, we will briefly discuss the notions of *risk* and *fundamental uncertainty*.

10.5.1 Risk

The notion of *risk* is crucial in economics and finance. Although it is defined in many different ways, most of these definitions implicitly connect the evaluation of

---

11. The probability $P(E|G)$ is called the *likelihood of E given G*, and it denotes the probability that the test will give a positive outcome given that Jones is guilty. As we have said already, however, this is not what we are after. By contrast, $P(G|E)$, which is the *posterior probability of G*, is the probability of Jones being guilty, given a positive outcome for the test.

risk to the evaluation of probability. For instance, in 1992 the London Royal Society defined risk as the probability "that a particular adverse event occurs during a stated period of time, or results from a particular challenge". In short, risk presupposes two dimensions: that of *utility* (how much, or how little, we benefit from a given situation) and that of *probability*. The Royal Society referred to probability "in the sense of statistical theory", emphasizing that it satisfies "all formal laws for combining probabilities". In the Stanford Encyclopedia of Philosophy's 2018 "Risk" entry, Sven Ove Hansson lists other definitions of this notion that have been given in the literature.[12] We can summarize them briefly as follows:

a. Risk = an undesirable event that may or may not occur.
b. Risk = the cause of an undesirable event that may or may not occur.
c. Risk = the probability of an undesirable event that may or may not occur.
d. Risk = the expected statistical value (the product of probability and some measure of its severity) of an undesirable event that may or may not occur.
e. Risk = the fact that a decision is made in conditions of known probability (*known unknowns*).

Roughly all definitions of risk presuppose the *statistical* application of probabilities, and therefore presuppose an "objective" view of the latter. Precisely for this reason it is important to distinguish between the *evaluation of risk* and the *perception of risk*. The first is based on statistical considerations, while the second is based on our degree of confidence (which is subjective) toward the given event (whether harmful or beneficial).

10.5.2 Fundamental uncertainty

We saw at § 10.1 that probability can be conceived as a measure of our *uncertainty* about events, and that under this view, the term "uncertainty" designates something subjective – the fact that we are not sure whether a given thing occurs or not. We have also seen (again at § 10.1) that the literature on the applications of probability also gives an "objective" sense to the term "uncertainty". In this sense, the occurrence of an event is uncertain if (i) it is not determined by predictable, known, or currently occurring factors, and (ii) it is not possible to make it the object of probabilistic evaluation, meaning that we have no rational justification for any attribution of a probabilistic value. Here, we will focus briefly on this sense of the term. In order to avoid confusion, we will refer to this sense of the term as to "fundamental uncertainty". This "objective" notion of uncertainty

---

12. Sven Ove Hansson, "Risk", The Stanford Encyclopedia of Philosophy, in Edward N. Zalta (ed.), https://plato.stanford.edu/entries/risk/.

is neatly exemplified by John Maynard Keynes, who referred to it as "uncertainty" *tout court*, and not as "fundamental uncertainty", as we do:

> By "uncertain" knowledge, let me explain, I do not mean merely to distinguish what is known for certain from what is only probable. The game of roulette is not subject, in this sense, to uncertainty; nor is the prospect of a Victory bond being drawn. Or, again, the expectation of life is only slightly uncertain. Even the weather is only moderately uncertain. The sense in which I am using the term is that in which the prospect of a European war is uncertain, or the price of copper and the rate of interest twenty years hence, or the obsolescence of a new invention, or the position of private wealth-owners in the social system in 1970. About these matters there is no scientific basis on which to form any calculable probability whatever. We simply do not know.[13]

The outbreak of a European war or the interest rate in twenty years' time depend on too many factors and the system is too complex for us to make sound estimates of their probability. Other terms to denote this "objective" type of uncertainty are "genuine uncertainty" and "profound uncertainty". We can therefore distinguish three types of situations:

a. *Known Knowns* (i.e. certainty). We know perfectly well the occurrence of events and their consequences. In these cases, we are totally certain that a given event has occurred, occurs or will occur.
b. *Known Unknowns* (i.e. probability applies). We have solid and defined grounds for evaluating the probability of a given event. For example, if I bet $100 on red in roulette, I am able to determine my odds of winning an additional $100 (48.64%) or, instead, losing my stake (51.36%). If tomorrow's weather forecast says it will be good and this forecast is 90% reliable, then I can determine the likelihood of my trip tomorrow being marred by bad weather (10%). In these situations, we know that we do not know that something will occur (or that it will not occur), but we also know how to measure the probability of this random event (or uncertain event, if we are thinking in subjective terms). Here, what *we do not know* (what is *unknown*) is that the event will (or will not) occur. What we do know (what is *known*) is its probability or the degree of confidence we have in it.
c. *Unknown Unknowns* (i.e. fundamental uncertainty). We have no reliable estimate of the probabilities of an event because our knowledge is too limited, and the system is too complex. For example, we do not know how to attribute a probability to the fact that we will have good weather in a certain area in 30 days or to the fact that the profits that a company will make in 20 years will be

---

13. John M. Keynes, *The General Theory and After: Defence and Development*, Macmillan, London 1973 ("The Collected Writings of John Maynard Keynes", vol. XIV), pp. 113-114.

greater than the current profits. Both the occurrence of the given event and its probability are unknown because it is *impossible* to define for the event a probabilistic value such that the assignment has a rational justification.

In some respects, the initial stages of the Covid-19 pandemic were characterized by a condition of fundamental uncertainty. The available data on the contagiousness of the virus and on the mortality rate were not sufficient to make rational decisions and only time will tell whether we overreacted or not. In this regard, Gerd Gigerenzer observed:

> No one knows where or how fast a new virus will spread. We cannot calculate the risks with confidence, and we will know only in hindsight whether we overreacted or underreacted. Given this uncertainty, how we respond to a viral outbreak is as crucial as the nature of the pathogen.[14]

These considerations confirm what we noticed at the beginning of this section, namely the close connection between probability and risk. Let us return to the definition of risk proposed by the Royal Society. According to this definition, risk is the probability "that a particular adverse event occurs during a stated period of time, or results from a particular challenge". Here, the Royal Society takes risk to involve "probability in the sense of statistical theory", which satisfies "all formal laws for combining probabilities". It is interesting to notice that this definition crosses paths with the three "categories" of situations just listed. According to this definition, in fact, we can speak of risk only in *Known Unknowns* situations, that is, in situations in which we can give a probability to the occurrence of an adverse event. Other definitions of risk envisage different intersections. For example, if we define risk as an undesirable event that may or may not occur, then we can speak of risks in such situations as well. There is a great variety of perspectives on this issue.

10.6 Conclusion

Let us recapitulate. We can intuitively characterize probability as a numerical description of how likely a given event is to occur. The notion of probability satisfies the rules established by probability theory, which is a branch of Mathematics. We have seen some of them, focusing in particular on Bayes' theorem, which helps us determine the *posterior* probability of an event, that is, the probability that one event will occur given that another has already occurred in the initial probabilistic space. This presupposes the notion of conditional probability, that is the prob-

---

14. Gerd Gigerenzer, *Why What Does Not Kill Us Makes Us Panic*, Project Syndicate, May 12, 2020.

ability that something occurs conditionally on something else occurring. Before discussing this basic theoretical point, we discussed three interpretations of the notion of probability: the classical interpretation, the frequentist interpretation, and the subjective interpretation. We also discussed their limits. Subsequently, we went through some problems in probabilistic reasoning, in particular problems to which we tend to give wrong answers. This has helped us understand our "fragility" as probabilistic reasoners – that is, the fact that in some situations, we prove prone to error in reasoning with probability. We have closed the chapter by discussing the notions of *risk* and *fundamental uncertainty*, which are both connected to the notion of probability.

11 | Reasoning by analogy

We close this volume with a further type of non-deductive reasoning: reasoning by analogy. We briefly illustrate its structure and discuss various purposes and applications of this reasoning; more precisely, we distinguish its cognitive and normative purposes; the latter are exemplified in legal and moral reasoning.

11.1 Introduction

Reasoning by analogy starts by identifying a *relevant similarity* between two or more entities and draws conclusions about *one* of the entities. One of the possible purposes of analogical reasoning is to broaden our knowledge, and in this case the entity about which a conclusion is drawn will be the lesser known of the two. In more technical words, reasoning by analogy projects a certain relational structure from one familiar domain (*source*) onto another similar domain (*target*), drawing conclusions about the latter.

Arguments by analogy are very common in everyday life. Since I liked the last two films of Christopher Nolan or the last two novels of Michel Houellebecq, I can reason by analogy and think that I will also like that director's next film or that writer's next novel; therefore, based on that, I can decide to go see the film or buy the novel. Since I like the haircut of a friend of mine and she goes to "Lidia, the magic touch" to have her hair cut, I can decide to go to the same hairdresser, thinking that the cut will be good for me too. Since I was satisfied with my latest mobile from the XYZ Microelectronics brand, I can decide to buy a mobile from the same brand, thinking that this will also satisfy me.

Like abduction and induction, analogy is a non-deductive and ampliative inference. It is therefore fallible but it can generate new knowledge, or rather it generates new knowledge when it is a good analogy. In this short chapter we will try to understand the structure of reasoning by analogy and some of its applications. In § 11.2 we will discuss the structure of arguments by analogy and the conditions under which they are considered (if their premises are true) good or not. We will then briefly look at three types of applications or purposes of reasoning by analogy. In § 11.3 we will consider arguments by analogy that have a cognitive purpose, in § 11.4 we will provide some applications of analogy to legal reasoning, and in § 11.5 some of its applications to moral reasoning.

11.2 Structure of arguments by analogy and evaluation criteria

The basic scheme of a reasoning by analogy is the following:

> Entity *a* has properties *P*, *Q*, *R* and *Z*.
> Entity *b* has properties *P*, *Q*, *R*.
> ―――――――――――――――――――――――
> Entity *b* has property *Z*.

For this reasoning to be appropriate, *P*, *Q* and *R* must be relevant properties, that is, properties connected in some way to *Z*. However, that things are really as the conclusion of a reasoning by analogy says is not sure; it is rather a hypothesis made by those who reason by analogy and this makes this kind of reasoning fallible.

In symbols, the scheme just presented can be rendered like this:

> $P(a)$, $Q(a)$, $R(a)$ and $Z(a)$
> $P(b)$, $Q(b)$, $R(b)$
> ―――――――――――――――――
> $Z(b)$

In this scheme, *a* is called the *source* of the reasoning, whereas *b* is the *target*. There is clearly a connection between this type of reasoning and the generalizations we talked about in Chapter 9. In generalizations we start out from the observation that a relevant portion of the members of a class have a certain property *Z* and we conclude that all members of that class have this property. Then, when faced with a member of that class that we have never met before, we can deduce that it has property *Z*. The first step of this reasoning (from some members to all members) is inductive, the second step (applying the generalization to a particular member) is deductive. For example, from the fact that the two mobile phones I owned from XYZ Microelectronics brand had satisfactory features and performance, I can induce that all the mobile phones of this brand have satisfactory features and performance and that, therefore, the next mobile of that brand that I buy will also have the same characteristics. Reasoning by analogy has similarities with this type of reasoning, but it goes directly from the fact that one or more individuals possess a certain property to the conclusion that another individual that resembles the former also has that same property, without going through the generalization to all individuals belonging to a certain class. The fact that the second individual resembles the first is indeed reduced to the fact that they share some properties that are somehow relevant to the conclusion that the second individual has property *Z* (see the argument scheme above).

There are various criteria by which we can evaluate reasoning by analogy:

1. *Relevance of similarities.* As we said, properties *P*, *Q*, *R* must be relevant for *Z*. If they are not, the argument by analogy becomes very weak. Suppose my friend Luisa was born on May 5 and she is very nice. I meet another woman who was also born on May 5 and I conclude that she must also be very nice.

Since the day a person is born seems *irrelevant* to the property of being nice, this reasoning is extremely weak.

2. *Number of similarities.* The higher the number of similarities between a and b, and thus the properties they share, the greater the probability that they share property Z. Suppose I want to buy a high-performance computer and suppose Paul's computer is the sort I want. Suppose that, on the web, there is an offer on a computer that has many features that Paul's computer has (processor, model and amount of memory, type of SSD, etc.). Clearly the greater the number of (*relevant*) features that the two computers have in common, the greater the likelihood that that computer is also high performance.

3. *Nature and number of disanalogies.* If the nature and number of similarities increase the strength of an argument by analogy, the nature and number of features that a and b do not share can weaken an argument by analogy. Suppose the last two Nolan's films I enjoyed were science fiction, but his next film will be historical. This difference undermines the conclusion that I will like Nolan's next film as well.

4. *Number of source individuals.* Suppose I not only enjoyed Nolan's last two films, but his last eight films. This reinforces the conclusion that I will also like his next film. Therefore, the higher the number of source individuals is, the stronger the conclusion.

5. *Specificity of the conclusion.* Consider the computer example again (point 2 above). If my conclusion is that my computer will have *exactly* the same performance as Paul's computer, then the argument is more easily falsifiable. A small deviation between the performance of the purchased computer and that of Paul's device is enough to falsify the conclusion. However, if we say that it is foreseeable that the purchased computer will have *more or less* the same performance as Paul's, the argument becomes stronger. So we can say that the more specific the conclusion is, the weaker the argument.

We will now turn to the different purposes that analogical reasoning can have.

11.3 Reasoning by analogy with cognitive purposes

In many cases we use reasoning by analogy in order to know, or rather predict, the properties of an object *that we do not know* in some respects. Reasoning by analogy does this by *projecting* onto such an object a characteristic that belongs to objects that are similar to it. In turn, such objects are similar to it because they share with the "unknown" object some properties that the latter has. Suppose I have seen *Inception* and *Tenet* by Nolan, and have not seen *Interstellar* (let us also assume, for convenience sake, that *Tenet* is a science fiction movie). In assessing whether there is a good chance that I will enjoy *Interstellar*, I reason as follows:

> I enjoyed *Inception* and *Tenet*, they are by Christopher Nolan, and they are science fiction movies.
> *Interstellar* is by Christopher Nolan, and is a science fiction movie.
> ─────────────────────────────
> I will enjoy *Interstellar*.

It is clear that this reasoning falls within the scheme seen above at § 11.2, and is therefore a reasoning by analogy. It is also clear that it is a type of reasoning that we apply frequently. Finally, it is clear what its *purpose* is: to *predict* a property (or lack thereof) of an object that I do not know in some respects. This is done by "projecting" onto it the properties of similar objects that I know, according to the scheme that shapes reasoning by analogy. The purpose of this reasoning is therefore *cognitive*.

11.4 Analogy in legal reasoning

Analogy plays an extremely important role in legal reasoning. In its normative sense, analogy has the purpose of extending, to a case *not expressly regulated*, the rules *expressly provided* for one or more cases with which the first has in common one or more *relevant properties* (by virtue of which the cases have a *relevant similarity*).

In this context the *target* is not a less known object, but an unregulated type of case. Similarly, the *source* is not a more known object, but a legally regulated case that bears a relevant similarity to the unregulated one. We will have a more concrete idea of what we mean by "legally regulated case" and "legally unregulated case" shortly when we discuss the American case *Adams* v. *New Jersey Steamboat Co.* (1896).

Reasoning by analogy in the legal field is especially relevant in English-speaking countries and, in general, in *common law* systems, i.e. those legal systems that are traditionally based on judicial precedents and build the rules of the legal system on them (the systems of *civil law*, whose great model is Roman law, in principle organize the norms into general codes so that the decisions of the judges do not create them, but apply them).[1] In a case law system, it is important for lawyers to establish the best possible analogy between a previous case, on which a judgment has already been made, and the case under consideration. Why is that? Because on the basis of the analogy the judge can "classify" the case in hand, and extend the type of decision made in similar cases in the past to it. In other words, if the analogy be-

---

1. However, it should be borne in mind that, in the last few decades, by virtue of the globalization process, the differences between legal systems (of common and civil law) have decreased; their distinction remains rather a matter of models.

tween the two cases is sufficient, the judge will be disposed to make the same decision that was made in the previous case. In any event, even in legal systems governed by legislative codes, reasoning by analogy ends up being used frequently since the laws are general and there are cases not regulated by such laws, or there are borderline cases that could be regulated one way or another. Furthermore, the general laws as well as the constitutional provisions are usually in need of interpretation. Their application to particular cases goes through interpretation. For example, the First Amendment to the United States Constitution guarantees freedom of speech and religious worship. Suppose a religious organization distributes leaflets to passersby. Since many passersby throw the leaflet on the ground after looking at it, we assume that the police, to avoid an excessive accumulation of garbage on the streets, order the organization to distribute the leaflets only in the vicinity of garbage cans. Suppose the organization objects that the police ordinance violates the First Amendment. In this case, it may not appear clear whether the case falls under First Amendment violations or those instances where the application of the First Amendment may be restricted by "significant government interests".

Arguments by *normative* analogy (or "for normative purposes") can be judged according to the criteria we saw in the previous section. The first criterion concerns the relevance of similarities. It is evident that relevance has an impact on legal arguments. For example, suppose an argument compares two events in which a building burned down and in which there have been victims. The comparison has no basis if in the first case the discussion concerns the compensation for damages owed by an insurance company and in the second case the charge of arson against a person held responsible for the fire. The number of similarities mentioned in the second criterion also has weight in legal arguments: the more similarities the two cases have, the more justified the conclusion that the two cases should be treated in the same way. Instead, and we are at the third criterion, the disanalogies between the two cases can weaken an argument. For example, if one case is based on a scam by a financial agent and another on a scam by a real estate agent, the differences between the two cases may be considered such that the two cases should be dealt with in the same way. As for the fourth criterion, the more similar cases have been treated in a certain way, the stronger the argument by analogy becomes. In other words, if a line of cases decided in a certain way is very numerous, the reasons for not following it must be very strong. Finally, as regards the fifth criterion, namely that of the specificity of the conclusion, imagine that, in a case of medical malpractice, a patient has been compensated with half a million dollars. An argument that another patient who was a victim of the same kind of medical malpractice should be compensated with a figure *close* to half a million is stronger than an argument that concludes that the patient must be compensated with *exactly* the same amount.

Legal reasoning by analogy is often elusive and it is difficult to find arguments that are not debatable. Analogies are often the result of creative hypotheses by lawyers and judges. Furthermore, one case may be similar to another in some respects and to a different case in other respects, and it is not always easy to identify which aspects are relevant or most relevant.

The American case *Adams* v. *New Jersey Steamboat Co.* (1896) offers an interesting example of this latter difficulty. It turned on a compensation claim from the passenger of a steamboat to the company providing the service, for a theft suffered by the passenger from the cabin of the boat assigned to him. The legal question was this: Can the theft on the steamboat be compensated without proving the fault of the service provider? There was no express regulation for the case. Was there, however, a rule applicable to the case by analogy? In fact, there were two, the regulation on hotel theft and the one on train theft, with the problem that the first was declared compensable without proof of fault and the second was not. So for a theft on a *steamboat* the question was posed in these terms: Is the relevant similarity the one with the hotel theft or the one with the train theft? Like a train, a steamboat is a means of transportation. Similarly to the hotel room, the cabin of a steamboat is a reserved space to which only the person providing the service (in addition of course to the customer) has access. If the relevant similarity was to the train, theft would not be compensated, but if it was to the hotel it would be compensated. Which analogical inference to draw?

The Court held that the relevant similarity was that between the boat and the hotel. The reasoning of the Court is not quite clear on this point, but it seems that the reason to award compensation for the hotel theft without proving the fault of the hotel owner was the protection of the trust placed by the customer in the service provider, with the consequence of making steamboat theft compensable in the same manner since the latter takes place in a reserved space in which customers can leave their assets trusting that no third party will have access to them. The reasoning could be schematically reconstructed as follows:

Hotel theft (*a*) must be compensated (*Q*).
Hotel theft (*a*) and steamboat theft (*b*) take place in reserved places to which only the service provider has access (*P*).

Steamboat theft (*b*) must be compensated (*Q*).

What inferential structure is involved in this scheme? *P* is a property deemed relevant that is common to the case of hotel theft and steamboat theft; it allows us to conclude that all cases involving such property must be ruled in the same way (in the absence of relevant exceptions or differences); then we must deduce the consequences for the specific case. Another property shared by *a* and *b*, that of occurring on a means of transportation, was considered irrelevant by the judges, or less relevant than *P*. But it is clear that a judicial decision like this can be debated.

Legal discussions often focus on the question of whether two cases are sufficiently similar to be ruled in the same way. Try thinking about examples like the following: if the Ku Klux Klan cannot be prevented from expressing their views, can a neo-Nazi party be prevented from expressing their own? And what about a fundamentalist organization? What is the relevant property to consider when deciding whether to extend or not the right to freedom of expression? Are there any relevant differences?

11.5 Analogy in moral reasoning

Reasoning by analogy is also used in the moral sphere. As in legal reasoning, the purpose of this argument is to look for relevant similarities between a case in which we have fairly firm moral intuitions and a case in which our intuitions are less solid. If the similarities are relevant and sufficient in number, we can conclude that we must apply the same moral norms to the case where our intuitions are less firm as we do to the case where our intuitions are clearer.

For example, suppose we ask ourselves whether a mother who takes excessive amounts of alcohol and drugs during pregnancy is morally (and possibly even legally) condemnable. It could be argued that those substances cause harm to the fetus and thus harm the unborn child throughout his or her future life. Since causing physical harm to another person is morally (and also legally) condemnable, the mother who takes drugs during pregnancy should be equally condemned.

However, one could reply that the fetus is not comparable to a person. For example, a person has the right to life, while, at least where abortion is legal, the mother can decide to terminate her pregnancy and, if the fetus has no right to existence, it has no rights in general. Therefore, the mother does not infringe any rights of the fetus if she takes such substances, because the fetus has no rights. However, it could be argued that a mother who decides to carry a pregnancy to term contracts duties, if not with the fetus, at least with the future unborn child, and that the infringement of the rights of such a future individual constitutes a moral fault on the part of the mother.

However, one could respond to this argument that it makes sense to contract duties only towards already existing subjects. One could, for example, refer to contracts: it makes sense to enter into a contract with an existing legal entity, but one cannot enter into a contract with a legal entity that does not exist (even if perhaps it will exist in the future). This could be countered by saying that this analogy does not hold because when the child is born he or she will have alcohol and drugs in his or her body and, as with any human beings who have alcohol and drugs in their body, we can ask ourselves where they come from and who is responsible for them. And it is clear in our scenario that these substances come from the mother, who is therefore behaving similarly to a drug dealer. One could try to counter that this analogy does not hold because it can only affect children who are born with traces of drugs and alcohol in their bodies but not children who are not born without such traces because the mother stopped taking those substances a few weeks before giving birth.

The discussion could obviously continue. However, we will stop here and note only that the arguments used in these replies and rejoinders are arguments by analogy. In fact, the arguments according to which a woman who takes substances during pregnancy is condemnable seek to establish analogies between the fetus and the human person, between the fetus and a legal entity with which a contract is established, and between the woman and a drug dealer. Since we are ready to morally condemn and those who cause physical harm to another person, those who do not respect a contract they have stipulated, and those who distribute drugs among the youngest, we should also be ready to condemn the woman who takes drugs during pregnancy, that is, we should be ready to extend the same moral judgment to this mother. The replies to these arguments, on the other hand, state that these analogies are not strong enough or rather that there are significant disanalogies between a person and a fetus, between a legal person and the unborn child, or between the mother and a drug dealer; these disanalogies would not allow us to treat the case of the woman and the fetus in the same way as the other cases in which we have solid moral intuitions. Hence, it is concluded that we should not extend the same moral judgment to women as we do in other cases. As in legal reasoning by analogy, also in moral reasoning by analogy the relevance of analogies and their number as well as the relevance of disanalogies and their number have a weight. The case of the woman who takes drugs while pregnant, like the case of the steamboat theft, shows that it is sometimes particularly difficult to say whether certain similarities are sufficient and sufficiently relevant.

If the similarities between the cases about which we have strong moral convictions and those about which we do not yet have them were clear and obvious, then we would not need to perform Critical Thinking and produce arguments by analogy. Similarly, if the analogies between a judicial precedent and the case under consideration were clear and obvious, lawyers and jurists would not have to produce arguments by analogy. It is the difficult cases that call for a greater critical discussion. Not only that, they also stimulate our creativity because finding analogies between different cases, as we have seen in the examples above, involves a creative effort on the part of the agents who argue to identify the best reasons for dealing with the cases to be decided. From this point of view, in such cases, Critical Thinking becomes a moral duty as well as a practical necessity.

11.6 Conclusion

Let us recapitulate. We have discussed reasonings by analogy, dwelling on their structure and the criteria on the basis of which we consider such reasonings good (if they have true premises) or not. We then discussed three different purposes or applications of reasoning by analogy, and in particular reasoning by analogy which has a cognitive purpose, and some applications of analogy to legal and moral reasoning.

Further reading

The following texts are useful for in depth analysis of the aspects dealt with in the various chapters of this volume.

Chapter 1: Rationality and cognitive biases

Dewey J., *How We Think*, DC Heath and Company, Boston 1910 (2nd ed. 1933).
Kahneman D., *Thinking, Fast and Slow*, Farrar, Straus and Giroux, New York 2011.
Kahneman D., Slovic P., Tversky A., *Judgment under Uncertainty: Heuristics and Biases*, Cambridge University Press, Cambridge 1982.
Osborne M., Rubinstein A., *A Course in Game Theory*, MIT Press, Cambridge (Mass.) 1994.
Thaler R., Sunstein C., *Nudge. Improving Decisions About Health, Wealth, and Happiness*, Yale University Press, New Haven 2008.
von Neumann J., Morgenstern O., *Theory of Games and Economic Behavior*, Princeton University Press, Princeton 1944.

Chapter 2: What is an argument?

Cohen M., Nagel E., *An Introduction to Logic and Scientific Method*, George Routledge and Sons, London 1934.
Copi I., Cohen C., McMahon K., *Introduction to Logic*, Pearson Education, Harlow 2014[15].
Perelman C., Olbrechts-Tyteca L., *Traité de l'argumentation. La nouvelle rhétorique*, Presses Universitaires de France, Paris 1958. Eng. trans. *The New Rhetoric: A Treatise on Argumentation*, University of Notre Dame Press, Notre Dame 1969.
Sinnot-Armstrong R., Fogelin W., *Understanding Arguments. An Introduction to Informal Logic*, Wadsworth, Belmont (Cal.) 2010[2].
Toulmin S., *The Uses of Argument*, Cambridge University Press, Cambridge 1958.
Walton D., Reed C., Macagno F., *Argumentation Schemes*, Cambridge University Press, Cambridge 2008.

Chapter 3: Rational discussion and the pyramid of disagreement

Hamblin C., *Fallacies*, Methuen, London 1970.
Lynch M., *Know-it-All Society. Truth and Arrogance in Political Culture*, Liveright Pub Corp, New York 2019.

van Eemeren F., Grootendorst R., *A Systematic Theory of Argumentation. The Pragma-Dialectical Approach*, Cambridge University Press, Cambridge 2003.

van Eemeren F., Grootendorst R., *Argumentation, Communication, and Fallacies*, Routledge, London 2016 (first ed. 1992).

Chapter 4: How to reply rationally to an argument

Walton D., *The New Dialectic. Conversational Contexts of Argument*, University of Toronto Press, Toronto 1998.

Walton D., *Dialog Theory for Critical Argumentation*, John Benjamins, Amsterdam 2007.

Chapter 5: Deductive arguments

Engel P., *La norme du vrai. Philosophie de la logique*, Gallimard, Paris 1989.

Kneale W., Kneale M., *The Development of Logic*, Clarendon Press, Oxford 1962.

Lemmon J., *Beginning Logic*, Thomas Nelson and Sons, London 1965.

Quine W.V.O., *Methods of Logic*, Harvard University Press, Cambridge (Mass) 1982^4.

Varzi A., Nolt J., Rohatyn D., *Logic*, McGraw-Hill Education, New York 2011^2.

Chapter 6: Conditional reasoning, I: The material conditional

Bennet J., *A Philosophical Guide to Conditionals*, Clarendon Press, Oxford 2003.

Ramsey F.P., *Foundations. Essays in Philosophy, Logic, Mathematics and Economics*, ed. by D.H. Mellor, Routledge & Kegan Paul, London 1978.

Woods M., *Conditionals*, Clarendon Press, Oxford 1997.

Chapter 7: Conditional reasoning, II: The counterfactual conditional

Goodman N., *Fact, Fiction and Forecast*, Harvard University Press, Cambridge (Mass.) 1983^4.

Hoerl C., McCormack T., Beck S. (eds.), *Understanding Counterfactuals, Understanding Causation: Issues in Philosophy and Psychology*, Oxford University Press, Oxford 2011.

Johnson-Laird P.N., *Mental Models*, Cambridge University Press, Cambridge 1983.

Lewis D., *Counterfactuals*, Blackwell, Oxford 1973.

Stalnaker R.C., "A Theory of Conditionals", in N. Rescher (ed.), *Studies in Logical Theory*, Basil Blackwell, Oxford 1968, pp. 98-112.

Chapter 8: Reasoning with explanatory hypotheses

Gabbay D., Woods J., *The Reach of Abduction. Insight and Trial*, Elsevier, Amsterdam 2005.

Hempel C., *Philosophy of Natural Science*, Prentice Hall, Upple Saddle River 1966.

Lipton P., *Inference to the Best Explanation*, Routledge, London 2004^2.

Mill J.S., *A System of Logic*, 1843 (ed. 1973 University of Toronto Press, Routledge & Kegan Paul, Toronto and London).
Walton D., *Abductive Reasoning*, University of Alabama Press, Tuscaloosa 2004.

Chapter 9: Statistical reasoning

Hacking I., *Logic of statistical inference*, Cambridge University Press, Cambridge 1976.
Hacking I., *An Introduction to Probability and Inductive Logic*, Cambridge University Press, Cambridge 2000.
Skyrms B., *Choice & Chance. Introduction to Inductive Logic*, Dickinson Publishing, Belmont 1966.

Chapter 10: Probability and probability biases

Aczel A.D., *Chance*, Thunder's Mouth Press, New York 2004.
Balducci A., Chiffi D., Curci F. (eds.), *Risk and Resilience*, Springer, Cham 2020.
De Finetti B., *Theory of Probability. A Critical Introductory Treatment*, Wiley, London 2017.
Knight F.H., *Risk, Uncertainty, and Profit*, Hart, Schaffner & Marx, Houghton Mifflin, Boston 1921.
Martelli A., *Models of Scenario Building and Planning: Facing Uncertainty and Complexity*, Springer, Berlin 2014.

Chapter 11: Reasoning by analogy

Holyoak K., Thagard P., *Mental Leaps. Analogy in Creative Thought*, The MIT Press, Cambridge (Mass.) 1995.
Levi E.H., *An Introduction to Legal Reasoning*, University of Chicago Press, Chicago 1949.
Schauer F., *Thinking Like a Lawyer. A New Introduction to Legal Reasoning*, Harvard University Press, Cambridge (Mass.) 2009.